Historical-Analytical Studies on Nature, Mind and Action

Volume 4

Editor-in-Chief
Professor Gyula Klima, Fordham University

Editors
Dr. Russell Wilcox, University of Navarra
Professor Henrik Lagerlund, University of Western Ontario
Professor Jonathan Jacobs, CUNY, John Jay College of Criminal Justice

Advisory Board
Dan Bonevac, University of Texas
Sarah Borden, Wheaton College
Edward Feser, Pasadena College
Jorge Garcia, University of Buffalo
William Jaworski, Fordham University
Joseph E. Davis, University of Virginia
Stephan Meier-Oeser, Academy of Sciences of Göttingen
José Ignacio Murillo, University of Navarra
Calvin Normore, UCLA
Penelope Rush, University of Tasmania
Jack Zupko, University of Alberta

Historical-Analytical Studies on Nature, Mind and Action provides a forum for integrative, multidisciplinary, analytic studies in the areas of philosophy of nature, philosophical anthropology, and the philosophy of mind and action in their social setting. Tackling these subject areas from both a historical and contemporary systematic perspective, this approach allows for various "paradigm-straddlers" to come together under a common umbrella. Digging down to the conceptual-historical roots of contemporary problems, one will inevitably find common strands which have since branched out into isolated disciplines. This series seeks to fill the void for studies that reach beyond their own strictly defined boundaries not only synchronically (reaching out to contemporary disciplines), but also diachronically, by investigating the unquestioned contemporary presumptions of their own discipline by taking a look at the historical development of those presumptions and the key concepts they involve. This series, providing a common forum for this sort of research in a wide range of disciplines, is designed to work against the well-known phenomenon of disciplinary isolation by seeking answers to our fundamental questions of the human condition: What is there? – What can we know about it? – What should we do about it? – indicated by the three key-words in the series title: Nature, Mind and Action. This series will publish monographs, edited volumes, revised doctoral theses and translations.

More information about this series at http://www.springer.com/series/11934

Ermanno Bencivenga

Theories of the Logos

Springer

Ermanno Bencivenga
University of California
Irvine, CA, USA

ISSN 2509-4793　　　　　　ISSN 2509-4807　(electronic)
Historical-Analytical Studies on Nature, Mind and Action
ISBN 978-3-319-63395-4　　　ISBN 978-3-319-63396-1　(eBook)
DOI 10.1007/978-3-319-63396-1

Library of Congress Control Number: 2017946848

© Springer International Publishing AG 2017
This work is subject to copyright. All rights are reserved by the Publisher, whether the whole or part of the material is concerned, specifically the rights of translation, reprinting, reuse of illustrations, recitation, broadcasting, reproduction on microfilms or in any other physical way, and transmission or information storage and retrieval, electronic adaptation, computer software, or by similar or dissimilar methodology now known or hereafter developed.
The use of general descriptive names, registered names, trademarks, service marks, etc. in this publication does not imply, even in the absence of a specific statement, that such names are exempt from the relevant protective laws and regulations and therefore free for general use.
The publisher, the authors and the editors are safe to assume that the advice and information in this book are believed to be true and accurate at the date of publication. Neither the publisher nor the authors or the editors give a warranty, express or implied, with respect to the material contained herein or for any errors or omissions that may have been made. The publisher remains neutral with regard to jurisdictional claims in published maps and institutional affiliations.

Printed on acid-free paper

This Springer imprint is published by Springer Nature
The registered company is Springer International Publishing AG
The registered company address is: Gewerbestrasse 11, 6330 Cham, Switzerland

Preface

I came of age as a logician and philosopher in the early 1970s, when one of the most popular games in town was adding to the plurality of logics—modal, epistemic, free, relevant, or what have you. After a while the proliferation of systems, with attending semantics and completeness proofs, became tedious and, having contributed my share to the industry, I checked out. But, wherever I went in the philosophical universe—whether I associated with Plato or Aristotle, with Descartes or Kant—I found logic present, not just as a tool that let me reason and argue about other things but as the very essence of philosophical work. I found that philosophers offered primarily, maybe even uniquely, different modes of reasoning about things. Ultimately, in fact, the things themselves did not matter and philosophical substance consisted of how you reasoned about them. And I found that some such modes were *really* different: as opposed to futile variations on how many multiple modalities you will allow, or what degree of logical omniscience you are prepared to allow to the epistemic subject, they really made you think of the world, and of yourself, and of your task in dealing with the world, with yourself, and with your interlocutors, in ways that made you, depending on the mode you chose, a different being, with a different understanding of her rational destiny.

I had a first epiphany of the monumental value of such confrontations in the early 1990s, when it finally dawned on me what Hegel was all about. I had known about his major theses since my days as a high-school student but previously could make no sense of them; so, instead of saying something stupid or making silly jokes, I had kept my mouth shut. When Hegel did make sense, it made enormous, if terrifying, sense, and the plurality of logics suddenly acquired thick, serious content. A third competitor emerged more slowly, through the 2000s, as I reflected on Anselm and Spinoza, and crystallized when I turned my attention to Bergson. There may be other competitors out there, for all I know; but it seems appropriate now to report on the status of the question as I have developed it in the last quarter century or so. In my last Kant book (2007), I say that "[r]eason is one and the same, yes, but as a war theater: it is traversed by an irremediable fissure, divided within itself, implicated in a perpetual internal confrontation" (p. 43). As it turns out, the fissure I talk about there is not the most basic one; in this book, I address one that traverses reason at an even more elementary level, before logic even gets off the ground (and that, as we will see, may be relevant to Kant interpretation as well—though such will not be my emphasis here). Just as Aristotelian being is originally divided (its division into categories is its very origin), so is logic—and there is not even anything here that would correspond to the primary sense in which, for Aristotle, substance is. All there is, as I will illustrate in what follows, is original multiplicity, original indeterminacy, and original play.

I thank Carlo Cellucci and Giuseppe Longo for their comments on an earlier draft of this book.

Irvine, CA, USA					Ermanno Bencivenga
May 2017

Contents

1   **Logic: A Contested Term** 1
2   **Analytic Logic** 5
      Interlude: A Conversation Between Two Analytic Neighbors 18
3   **Dialectical Logic** 25
      Interlude: A Conversation Between Two Dialectical Neighbors 36
4   **Oceanic Logic** 43
      Interlude: A Conversation Between Two Oceanic Neighbors 55
5   **Necessity** 61
6   **Truth** 77
7   **Negation** 91
8   **Infinity** 107
9   **Mathematics** 119
10   **Texts** 131
11   **The Play of Logics** 141

Coda: A Conversation Between Two Real Neighbors 151

Bibliography 161

Chapter 1
Logic: A Contested Term

Abstract Logic is typically characterized as a (or the) theory of inference; but that characterization is a biased and restrictive one. Once we understand a logic, instead, as a theory of the *logos*—that is, of meaningful discourse: of what makes discourse the vehicle of meaning—we open up a field of possible ways of reasoning and arguing that do not reduce to inferences. It is within this field that the present book moves.

If we look up the word "logic" in a dictionary, philosophical or otherwise, we are most likely to see it defined by some variant of "theory of inference"[1]; if we go just slightly below the surface of this characterization, we will see that what the theory is supposed to do is specify parameters by which an inference can be judged "valid"—that is, such that some part of it (the conclusion) "follows" from some other parts (the premises). The word "reasoning" is also going to be thrown in there but in such a way that reasoning is uncontroversially identified with *inferential* reasoning and *good* reasoning identified with running valid inferences. After that, we are up and running, talking about possible worlds and formal sentential patterns and rules for validly inferring some such patterns from some others. But, after that, most of what matters about logic has already been decided—much as happens at philosophy talks, when the speaker, "before beginning in earnest," asks the audience, amicably, to settle "a few terminological matters" and, if the audience agrees to settle them as he proposes, the claims to be made in his paper have already been established: it is just a matter of extracting them from the agreement the audience was too quick to come to.

The space within which the present book intends to move can only be gained, hence the book can only have a chance to unfold, if we resist this knee-jerk reaction: if we pause and ask ourselves what logic could be *other than* a theory of inference and reasoning *other than* a series of inferences. So let us consider a case in which I am facing an opponent and *arguing* with him—in the language to be used here,

---

[1] In the *Merriam-Webster* dictionary the first definition of "logic" (more about multiple definitions below) is "a science that deals with the principles and criteria of validity of inference and demonstration." In *The Oxford Dictionary of Philosophy* "logic" is defined as "The general science of inference."

"argument" and its cognates will be more neutral terms than "inference" and its cognates and refer to all kinds of rational linguistic strategies, not always inferential, by which we attempt to prevail over an opponent. Say that A is a libertarian and is defending a minimal state, where everyone is left alone to fend for herself and social welfare is reduced to a bare minimum. If I disagree with him, I could try to prove him wrong, and my best chance for doing so would be by examining his various tenets in detail and arguing that there are contradictions arising among some of them: since contradictory statements cannot both be true, A's position (if I were successful) would thus fall of its own weight, and he himself, insofar as he sticks to it, would be shown to be an irrational person. That would be a use of logic (and argument, and reasoning) based on inference: I would infer a contradiction from his premises and thereby refute them.

But I could also proceed otherwise. I could, for example, bring up for discussion this notion of liberty that A cares so much about. I could say that I firmly agree with him on the value of liberty—even on its foundational value for what it is to be human. But, I would continue, we need to get deeper and clearer about the substance of liberty: about what "liberty" really means. Can a person be said to have liberty, to be free, when the social environment in which she was born and raised, and where she leads her entire life, is one that systematically stunts her growth, crushes her promise, and denies her any opportunity for flourishing? Shouldn't a libertarian like A, by simply working out the notion of liberty, become convinced that the whole community would be freer if it gave every one of its members all that is needed for them truly to express themselves, truly to make choices, truly to work at their dreams and thus make apparent to all of us what they—themselves and their dreams—are worth?

Though I would personally sympathize with this line of discourse, my personal sympathies are not at issue here (and, indeed, I will soon evoke a less sympathetic picture of the whole performance). What is at issue—the issue I want to raise—is rather: Couldn't this line of discourse be regarded as a form *of reasoning*? Wouldn't I, by engaging in it, be using my reason and making an appeal to the reason of my interlocutor, in order to prevail over him—or at least over those views he currently espouses? But there is no inference here; in fact, though I am certainly trying to have my opponent see the light (of reason), I am not even trying to prove him wrong! What I am doing, instead, is subtly appropriating and incorporating his position: presenting it as something that is not so much to be refuted as overcome—as an early chapter in a story that inevitably, by working itself out, will arrive at the position *I now* have. Anyone who endorses some variety of paternalism (here comes the less sympathetic part) reasons and argues that way: children, women, non-Caucasians, and "savages" are just fine in their beliefs and practices—except that those beliefs and practices are somewhat "primitive" and will have to evolve into the more mature stage represented by ourselves. Into *their* more mature stage, to be explicit, for, by evolving that way, they will not relinquish their identity but make it express itself more fully and become more of what it is. Children and savages *are us*, though they have not realized it yet—and they will be more themselves, truer forms of themselves, the more it becomes apparent (to themselves and others) that they are us.

1 Logic: A Contested Term

Nor does it end there. A minimal state has minimal tasks, but it does have *some*. It has one task, primarily: protecting the citizens from harm. So I ask my interlocutor to consider an obvious case of harm: a person attacked in the street and punched in the face. You would want to avoid that sort of occurrence, wouldn't you? And, in case it had already taken place, you would want to apprehend the perpetrator, lock him up, and take appropriate steps so that he does not offend again, right? Well, what about someone (not punching but) insulting another, calling her names, using racial slurs on her? Is that harm? Wouldn't you want to do something about it? What about someone using his power and wealth to deny someone else food or shelter or to solicit sexual favors from her? Is that harm? What about someone being destitute (because the community takes no care of her) and *forced* to provide sexual favors in order to survive? Is that harm? Wouldn't you want to address it? Or where would you draw the line between the harm you recognize and address and the one you do not? On what basis do you decide to draw the line there? What justifies that decision?

Once again, the above is not an inference. If anything, it demands an inference from the opponent, one that would legitimize a particular division between harm and non-harm; and it stands ready to contest any such inference and division by judging it groundless—a case of drawing the conclusion one wants from arbitrary, self-serving premises. But there is no denying that reason is being used in this process and that by making such moves I would be *arguing* and possibly creating trouble for my interlocutor—and doing so precisely insofar as his reason would be queried by mine and might find itself unable to respond to it.

One could insist that these kinds of performances, however instructive or effective, do not pertain to logic: that logic is just a theory of inference, and hence, insofar as no inference is made here, logic does not belong in the proceedings. To those who decide to take this route, I have nothing more to offer; their very understanding of the word "logic" makes it impossible for them to reason with people who share no basic premises with them. For people like that, it must be left at the stage of "he says; she says." But there may be others who will agree that the above are cases of reasoning, however different from, say, a step in the proof of the Pythagorean theorem; and with those others I can have a conversation. Where the first subject to be addressed is: if logic is not a theory of inference, but some more general endeavor of which a theory of inference is a particular case, what is logic then?

My answer to this question is based on etymology. Logic—or, more precisely, *a* logic—is a theory of the *logos*, of meaningful discourse: of a discourse that makes sense, that manifests reason. So, before a logic goes any further—specifically, before it licenses certain forms of confronting an opponent and prevailing over him—it must tell us what a *meaning* is and how a piece of language (a *word*, for short) acquires it. There are, it turns out, at least three major ways (intimated in the examples above) in which a theory can do that; in this book, I will first introduce them (in the next three chapters) in a summary way, and then I will further develop their similarities and differences by having them confront specific (and fundamental) issues.

Chapter 2
Analytic Logic

Abstract The theory of the *logos* that is typically considered the only existing logic and will be called here *analytic* logic is based on Aristotle's work; in recent times it has gone through important developments, foremost among them a new and more satisfactory treatment of relations due to Frege. But what remains true of it is that language and meaning are organized by contraries: predicates that cannot be both true of the same thing (at the same time) and, by extension, sentences that cannot both be true (at the same time). Given contraries, words can be given *definitive* definitions, and arguments can reach *conclusive* conclusions—if they are judged to be correct, they are never going to be revised. Inferences are arguments of this sort: arguments from premises to conclusions. Because of their conclusive character, inferences can be formalized: one can abstract from them the minimum of content that makes it possible for the conclusions to be established and be assured that, whenever that same minimum of content is identified in an inference, a corresponding conclusion will be warranted.

Analytic logic has its most natural application to mathematical entities and to other entities (e.g., Platonic ideas) that philosophers have modeled on mathematical ones. Its most important anomalies (i.e., signs of imperfect adaptation between the theory and its subject matter) are (a) ambiguous words, which are associated with multiple (contrary) definitions; (b) vague predicates, which gradually merge with their opposites, thus contesting the very notion of a contrary; and (c) spacetime continuants, which constantly change all their identifying properties (into some of their contraries) while remaining self-identical. That these are anomalies, rather than problems to be eventually resolved within the theory by applying sufficient ingenuity, is suggested by their intractable nature; but accepting this view of them cannot follow from simple despair—it must be based on being presented with alternative theories of the same subject matter.

Aristotle has been universally credited with inventing logic, occasionally in extravagant ways: Kant, for one, thought that he not only initiated the discipline but also brought it to an end.[1] And the logic universally attributed to him is a theory of

---

[1] "That from the earliest times *logic* has traveled this secure course can be seen from the fact that since the time of Aristotle it has not had to go a single step backwards, unless we count the aboli-

inference, which later scholars expanded and systematized, translated into mathematical language, and codified into formal calculi. But only one of the six treatises comprising the *Organon*, Aristotle's collection of logical works, deals with inference and inferential patterns: the *Prior Analytics*. What else is contained in the *Organon* includes (among other things) a review of the basic furniture of the universe (in the *Categories*); a distinction between spoken and written symbols, which vary across languages, and the mental experiences those symbols signify, which are the same for everyone (in the *De interpretatione*); and a theory of definition (in the *Posterior Analytics*). Which is consistent with my view that Aristotle's conception of logic as a theory of inference is an outgrowth of his conception of how words get attached to things—how they acquire meanings.

What mediates word-world connections, in the Aristotelian framework, is definitions, which are built by genus and differentia. Say that you want to define *human*. You start with the original genus (or category) *substance*, of which there is no higher (*being* is not a genus[2]), and then proceed stepwise down a tree of Porphyry (actually, to me it looks more like roots). You face a first branching of the tree when you are given a choice between *material* and *immaterial*, you take the first course and run into a new choice between *animate* and *inanimate*; you take the first course and run into a new choice between *self-moving* and *non-self-moving*. Again you choose the former and label that node *animal*; further down, you must select either *rational* or *irrational*. You pick *rational* and hit your target: the things to be called *human* will have to be found among the genus *animal*, and what makes the difference between them and other animals is that they are rational—all other animals are brutes. To put it otherwise, the meaning of *human* is specified by its definition as *rational animal* and thereby *analyzed* into a collection of traits, which number two in this shorthand version but in fact are seen to number more if we remember what the genus *animal* and the differentia *rational* (which will be found in another tree, whose original genus is the category of *quality*) themselves mean.

The analysis I sketched above still has, at this stage of the game, loose, unpretentious features. One could imagine that the meaning of a word be broken down into various elements and then these elements be recombined into the meaning of some other word, or different elements be combined into another meaning (or even the same one) of the same word. But this is not how it works here, because of a second crucial feature of the Aristotelian framework: all steps taken in a definitional process are irreversible; the analysis a definition provides of a word is not a playful combinatory exercise; it is a postmortem. It is this feature more than any other that qualifies the framework and the resulting logic as *analytic*. For *the* analysis you

---

tion of a few dispensable subtleties or the more distinct determination of its presentation, which improvements belong more to the elegance than to the security of that science. What is further remarkable about logic is that until now it has also been unable to take a single step forward, and therefore seems to all appearance to be finished and complete" (1998, Bviii). As I mentioned in the preface, there is more to be said about the relation of Kant to logic(s) than this passage suggests; I will get to it in due course.

[2] See, for example, *Metaphysics* 998b.

provided of the meaning of *human* (notice the definite article) is all there will ever be to say about that meaning.

Return to the choice between *animate* and *inanimate*. These two qualities (or *predicates*, to use a term that has more currency today) are *contraries*: they *cannot* both be in the same substance at the same time (if seen as predicates, cannot both be true of a substance at the same time). Hence, if you chose *animate*, the option of later choosing *inanimate* is foreclosed; your choice is irrevocable. In addition to being contraries, *animate* and *inanimate* are also *contradictories*—they also cannot both *not* be in a substance at the same time—and contradictories tend to monopolize the relevant press; but it is their nature as contraries that is crucial for the point I am making now. If after leaving *animal* behind you were to face a node that required a choice between *mammal*, *bird*, *fish*, *reptile*, and *insect*, and decided to go the way of *mammal* in looking for *human*, your choice would be no less irrevocable for the fact that *mammal* is no contradictory but (only) a contrary of *bird* (as is of *fish*, *reptile*, and *insect*). What would make it irrevocable is the simple fact that the quality of being a mammal cannot coexist in a substance with the quality of being a bird (or a fish, a reptile, or an insect)—regardless of whether or not everything must also be one or the other.

Contraries structure the Aristotelian universe, and contrary predicates structure the Aristotelian discourse. This correspondence would make for an interesting discussion of how far our language determines our world; but such discussion, attractive as it is to a Kantian like myself, will have to wait for another day.[3] For the moment, I take stock of the correspondence and put it to use: what is said on the side of words is often extended to the side of things; the relations among things often echo the relations among words. And, with that in mind, I focus on one element of the definition of contraries: on the modality "cannot." This modality makes us go beyond mere description in our account of the data (later I will discuss what the data are here): it justifies lawlike statements ("No birds are reptiles") and sustains counterfactuals ("If I were a bird, I would not be a reptile"). But how is it itself to be understood? By appealing to the intertranslatability of basic modal notions, we could say that it introduces a form of *necessity* ("Necessarily, no birds are reptiles"); but, again, what form is that? What force does the adverb "necessarily" have in "Necessarily, no birds are reptiles"?

The necessity in question here is nothing other than the irrevocability I already mentioned: once a thing *a* has been identified as a bird, it is forever excluded from the range of reptiles. A sentence like "*a* is a reptile" will never have to be considered again: we can cross it out and retire it for good. Which makes for an attractive program in the sciences: they are to carve nature at its joints; hence to establish what falls on either side of any particular joint and, having thus irremediably *fallen* there, can never again be regarded as able to negotiate the divide between the two sides.

Once an appropriate definition of *human* is obtained by a judicious selection among contraries, humans will continue to face similar choices among other

---

[3] A weaker version of the thesis alluded to here (our language determines how we think of the world) is relevant to what follows and will be addressed in Chap. 11.

contraries. Sometimes collectively: they will have to go one way or the other, *but not both*, between *mortal* and *immortal*. Sometimes individually: each of them will have to go one way or the other, *but not all*, among *white*, *black*, and who knows what other color. As these choices are made, the notion of meaning will expand from words to sentences: we will determine meanings for, say,

(1) All humans are mortal.
(2) Bruce Springsteen is white.

And the same necessity—irrevocability, that is—will continue to govern our understanding of these sentences: once (1) is asserted, the mortality of humans will never have to be revisited. By asserting (1) we have carved (human) nature at one of its joints, so this issue is closed.

Here is where inference comes in. Consider:

(1) All humans are mortal.
(3) Socrates is human.
(4) Therefore, Socrates is mortal.

(4), we are told, follows from (1) and (3), which means: necessarily, if (1) and (3) are both true, then (4) is true. Part of what this says is trivial, and it is the part that regularly receives the most emphasis: what is true collectively of all men is also true distributively of every one of them (including Socrates). What is not so trivial, indeed constitutes the distinguishing character of this logic—what properly characterizes it as *inferential* logic—is that this argument justifies, indeed, a *conclusion*: there is something conclusive and definitive about it. Since sentences have contraries, too, we can be sure that a contrary of (4) like

(5) Socrates is immortal.

(which is also a contradictory of (4); but, again, that is irrelevant) is once and for all out of bounds. In a similar way, the inference

(6) All white things are not black.
(2) Bruce Springsteen is white.
(7) Therefore, Bruce Springsteen is not black.

will establish once and for all the untenability of

(8) Bruce Springsteen is black.

Let us get deeper into this. Since (4) is obtained as the conclusion of an inference, it will never be revised, I said. There is nothing more to be learned about its truth—or falsity, if that were what the inference established. But its truth is a simple function of its relation to the premises of the argument, *that is*, of the relations of subordination those premises assert between *human* and *mortal* and between Socrates and *human*. Those relations contain all the information needed to determine the validity of the argument from (1) and (3) to (4); hence any other argument in which the same relations were in play would be valid for exactly the same reasons and would establish the definitive, conclusive truth of its conclusion (assuming that the premises are true) in exactly the same way. Consider one such argument:

(9) All Athenians are Greek.
(10) Plato is Athenian.
(11) Therefore, Plato is Greek.

(9)–(11) bring up a whole other set of characters: an individual who is not Socrates and qualities that are not *human* and *mortal*. But for the validity of (9)–(11), and for the truth of (11) if the premises are true, none of that matters: the premises assert the same relations of subordination between the relevant terms, *and that is all*—we do not expect to ever learn anything about Plato that would force us to revise this conclusion. The differences among the various individuals and qualities involved in the two arguments, if anything, could be distracting from what does all the logical work in *both* of them; so it is convenient and advisable to elide them and represent them by simple, unanalyzable symbols—symbols that, not being analyzable, carry no meaning. The same symbols for the terms in the two arguments that enter those crucial relations in the same way, to obtain something like the following:

(12) All As are B.
(13) s is A.
(14) s is B.

It is by this route that we arrive at another major feature of analytic logic: its being *formal*. That a logic be formal is often accounted for by saying that it disregards the content of sentences and concentrates on their form only. But this is confused: copulas and quantifiers (or conjunctions or adverbs—the parts of speech that analytic logic tends to focus on, that it even sometimes considers specifically *logical* words[4]) are as much part of the content of a sentence as nouns and adjectives are. So the account must be modified: a logic is formal if it disregards *some* of the content of our sentences and concentrates on some other such content only—if it finds it legitimate and sufficient for its purposes to disregard some sentential content and to pay no heed to it. The letting go of *any* of our meaning when making logical points is enough to qualify our logic as formal, much as the letting go of any of our personality within a relationship with someone else is enough to qualify that relation as formal.

(A) logic is a normative theory: it does not concern itself with how people understand language or argue with one another; it states how people *must* understand language and argue with one another. It imposes rationality on us. So a modal

---

[4] A whole diatribe has been running through the literature concerning what logical words are: a fake issue that has wasted people's time. For the only sense that can be made of calling something a logical word is: a word whose logic is being studied (which makes the term dependent on the contingencies of what is being studied logically). Fortunately, as more and more words and phrases have become the subject of logical scrutiny, this issue has receded into the background, and it no longer has the currency it enjoyed at the time of Quine—that notorious issuer of draconian, unenforceable edicts. On this topic, see my (1999).

formulation of its principles is intrinsic to it.⁵ In analytic, Aristotelian logic, the fundamental modality on which every other is based is (as I pointed out earlier) the "cannot" contained in the definition of contraries: it makes every other definition the last word on what something means (*definitive* on it—pun intended, as with the conclusiveness of a conclusion), and it allows arguments to reach unrevisable outcomes—to be inferences. A lot has changed in analytic logic since Aristotle; most importantly, Gottlob Frege added a much more adequate treatment of relations to it. But this essential aspect of it has remained untouched: the conclusion of an argument like

(15) No man is his own father.
(16) John is Paul's father.
(17) Therefore, John is not identical with Paul.

is just as irrevocable as either (4) or (7), and it is just as irrevocable what the meaning of the relation "being father of" (as opposed to the property "being Paul's father," which is all that Aristotle could do justice to), when properly analyzed, will amount to.⁶

A theory is supposed to "save"—that is, explain—some range of phenomena and to be answerable to those phenomena: to have currency insofar as it does explain them. In the case of a logic, the phenomena to be explained, and to be answerable to, border on the evanescent: they are not, I said, common patterns in the actual use of a language (which could be determined by empirical, statistical studies, but may well be fallacious) as rather *intuitions* by competent speakers on how language is *properly* used (i.e., *normative* intuitions, analogous to those a grammar is supposed to explain—since what we could call the *logicality* manifested by such intuitions is analogous to the *grammaticality* posited by Noam Chomsky).⁷ The near-evanescent

---

⁵The modality pertinent to a normative theory is *deontic*, as opposed to *alethic*: often what *must* happen does *not* happen (and "should" is then a more common term in it than "must"). So endorsing a logic is compatible with admitting that people often think, reason, and argue in violation of it, much as endorsing a theory of justice is compatible with admitting that people often do what the theory regards as unjust.

⁶If anything, the primacy of contraries has become even clearer in post-Fregean (analytic) logic. Consider, for example, the following: all (infinitely many) sentential connectives can be defined in terms of the single binary connective "either not A or not B" (the so-called Sheffer stroke), which states A and B to be contrary—or inconsistent with one another. Brandom (1979) identifies a variant of the Sheffer stroke that makes it possible to define both connectives and quantifiers. Strawson (1952), in addressing logical appraisal, selects the "notion of inconsistency or self-contradiction for detailed discussion" (p. 2).

⁷That logic only be answerable to normative intuitions is how I understand its traditional a priori status: its claims are independent of any empirical data and will not be impacted by them. It is an empirical fact that someone *has* those intuitions, but there is no empirical content to them: their normative character insulates them from the need to receive empirical confirmation, though there can still be an empirical debate among people empirically committed to the same logic—about what their intuitions in fact tell them. (So I reject a stronger notion of logic being a priori: that its correctness holds objectively, whatever people's intuitions might be; and what I said on its setting up a program for the sciences must be understood in terms of what our intuitions warrant, not what the world objectively is.) On the other hand, if I had an intuition that now it is raining in Moscow,

nature of the data makes the adaptation of a logic to its field of application especially plastic and makes that field subject to logical regimentation,[8] and yet, it continues to hold for a logic, as for any other theory, that some of the phenomena in its range look like *natural cases*—cases that the theory fits perfectly—while others look like *anomalies*—recalcitrant ones, constant sources of confusion and irritation.

The most obvious natural case for analytic logic is mathematics; hence it was not by chance that Galileo recommended the study of mathematical proofs, not of logic treatises, to become proficient in logic[9] (and I referred to the Pythagorean theorem above)—as always, practice makes perfect; and the logic he was referring to was the analytic one. A number is necessarily all it is, and so is a geometrical figure, in the sense of "necessarily" that is operative in the present context: when a theorem is proved that establishes some property of a number, or of a geometrical figure, or some relation among numbers or among geometrical figures, the theorem becomes part of the permanent depository of mathematical truths, possibly to be revised in the future by a generalization or a refining of its claim but never to be *denied*. A denial of a mathematical theorem would amount to proving another, distinct theorem, to the effect that the previous one and its (alleged) proof *were wrong*.

Not surprisingly, then, mathematical objects and truths have recurrently, and authoritatively, functioned as archetypes for all respectable objects and truths. In the most extreme example of this deference to analytic finality, Platonic forms displayed the same immutable character as mathematical entities, though hardly the same fertile capacity to generate a rich and varied landscape (and Aristotle's logic proved to be a vindication of his teacher's logical intuitions—another topic I will leave aside here). Of a similar ilk were thought to be concepts, essences, Fregean senses, Husserlian noemata, and other assorted (and exalted) specimens that have constantly populated philosophical texts—though with most of them the resemblance wore thin because their champions were not prepared, as Plato was, to declare them the only existing objects and truths about them the only existing truths;

---

that *descriptive* intuition would have to be compared with the empirical data before it could be trusted. Whether there are any descriptive intuitions (say, the intuition that every event has a cause) that are as insulated from the empirical data as the normative ones *and that can be trusted* is, of course, the problem of whether synthetic a priori judgments are possible.

[8] Think of Russell's theory of descriptions as a prime example of this kind of regimentation. And think of what the implications are of it having been called "a paradigm of philosophy" (by Ramsey 1990, p. 1).

[9] "Logic, as it is generally understood, is the organ with which we philosophize. But just as it may be possible for a craftsman to excel in making organs and yet not know how to play them, so one might be a great logician and still be inexpert in making use of logic. Thus we have many people who theoretically understand the whole art of poetry and yet are inept at composing mere quatrains; others enjoy all the precepts of da Vinci and yet do not know how to paint a stool. Playing the organ is taught not by those who make organs, but by those who know how to play them; poetry is learned by continual reading of the poets; painting is acquired by continual painting and designing; the art of proof, by the reading of books filled with demonstrations—and these are exclusively mathematical works, not logical ones" (2001, pp. 39–40).

therefore, some accommodation had to be found, with great difficulty, between them and the less-than-respectable entities we are going to be considering shortly.

Turn to the anomalies now. To begin with, ordinary words are *ambiguous*: a vast majority of them (indeed, all of them except "technical" ones) have multiple meanings. Just open up the dictionary again and look up any word (*including* "logic"): what you are going to find, in most cases, is a paragraph of variable length, neatly organized by numbers and letters (or vice versa) which indicate divisions and subdivisions of that word's meaning. Each subdivision is intended to report an Aristotelian definition of the word, by genus and differentia; but, clearly, any such definition is far from settling, once and for all, the word's contribution to meaningful discourse. Aristotle, who saw the issue from the opposite side and, instead of ambiguity, spoke of *homonymy* (i.e., of different things having the same name), was extremely worried about this phenomenon, so much so that the very first paragraph of his entire corpus (the first, brief paragraph of the *Categories*) is a definition of homonymy[10] and, from then on, homonymy surfaces constantly in his works, together with mention of measures to be taken in order to defuse it. Not surprisingly, for how are you going to make any sense if you are never sure of how what you say will be understood, or *should* be understood?

The worry has not gone away. Occasionally, it blows out of proportion when a philosopher realizes that what is true of God and humans (that the same words applied to the one and the others—words like "knowledge," "will," or "goodness"—have different, even incommensurable, meanings) may also be true of individual humans: for each of them, a specific idiolect may be needed to describe her experience, different from, and incommensurable with, the idiolect relevant to anyone else. In less extraordinary circumstances, the worry flies under the radar, and people accept as a fact of life—a bothersome one, but still one they can (or need to, as they have no choice) live with—that those carriers of meanings which words are supposed to be regularly splinter into identical copies of themselves, and we must think of them as indexed by numbers or letters (or both, as in the dictionary), thus bringing out the ontological multiplicity hidden under their apparent, misleading sameness. Then the worry heats up when someone tries his hand at a scientific or philosophical pursuit that would benefit from all the precision supposedly granted by analytic logic, finds himself unable to trust words that mean different things for different people, and feels inclined to introduce (I suggested above) new *technical* (i.e., made-up) words whose meanings he legislates by exercising his own authority and thereby providing a definition that no one is going to be in a position to mess up with.[11] But finality is thus acquired on the cheap; and the problem is left of what

---

[10] "When things have only a name in common and the definition of being which corresponds to the name is different, they are called *homonymous*. Thus, for example, both a man and a picture are animals. These have only a name in common and the definition of being which corresponds to the name is different; for if one is to say what being an animal is for each of them, one will give two distinct definitions" (1a).

[11] The introduction of new words is often justified by saying that what they refer to is also new and that the use of existing words might generate confusion with the existing meanings of those words. In the next chapter, we will see that the very notion of what is new here is controversial: depending

relation (if any) there is between this new vocabulary established by an act of will and the one everybody else continues to use in her everyday life—and the everyday world the latter vocabulary refers to. The problem, that is, of how that scientific or philosophical pursuit, described as it is in a language that is wholly internal to it, might be relevant to anything we care about.

The ambiguity of ordinary discourse fits the role of an anomaly: it signals an imperfect adaptation between the Aristotelian theory of definition and the phenomena the theory is supposed to explain—people's logical intuitions about how words are properly understood and used. And it is an imperfection that will not go away, despite the concerted efforts repeatedly made to disarm it, which distinguishes the case of an anomaly from that of a *problem* that might also arise in a theory. A problem can be solved, and then it is no longer a problem; an anomaly can never be solved, it is always looming on the horizon and pointing to an inadequacy; the only successful way of dealing with it is by looking at things differently and moving to a different theory. (The precession of the moon's perigee was a problem in Newtonian mechanics and was eventually solved; the precession of Mercury's perihelion was an anomaly for it and was only "solved" by moving to relativistic mechanics.) But the commitment to analytic logic runs strong and deep; hence it is hard to dislodge it by just pointing at a particular, however insistent, source of nuisance. Our best bet, if we want to make people interested in looking at other options, is to accumulate various sources of nuisance—and then of course offer those other options. So I will list here two other such sources, which give rise to (different kinds of) paradoxes: to statements beyond belief. Ambiguous words, of course, can also be used to generate paradoxes, for example, the one brought forth by the following argument:

(18) Sharp things cut.
(19) The note A# is sharp.
(20) Therefore, the note A# cuts.

But people think that they can easily see through this *quaternio terminorum* and the "logical illusion" (Kant's term) it generates, and that once they see the fallacy involved in it (Kant himself would claim), they are not going to be fooled again (the illusion is forever dispelled).[12] The paradoxes I am going to bring out now, on the other hand, are not so comfortably resolved.

Begin with something that looks marginal and is—but no less troubling for being so. When coming down the definitional tree of Porphyry to find *human*, we hit a node that required a choice between *self-moving* and *non-self-moving*. We chose the first course and labeled the outcome *animal*; if we had chosen differently, we would

---

on what logic you adopt, it might mean something radically distinct from whatever existed previously or, instead, a new phase of a continuing process (in which case the use of a new word would itself be confusing).

[12] "Logical illusion, which consists in the mere imitation of the form of reason (the illusion of fallacious inferences) arises solely from a failure of attentiveness to the logical rule. Hence as soon as that attentiveness is focused on the case before us, logical illusion entirely disappears" (1998, A296–97 B353).

have labeled it *plant*. And, either way, our choice would have been a definitive one: *animal* and *plant* are contrary; no animal is a plant. But there are, indeed, margins to these two contraries; and Aristotle was aware of them. He was aware, for example, that it is hard to decide what we should call sea anemones, as they resemble plants on one hand and animals on another.[13] We might add other examples: what should we call comatose patients, or fetuses? Here, too, some might try to minimize the issue and the worry that goes with it: we know that lions are animals and that sloths are (!?), and we know that oaks are trees. What we know along those lines is enough to have us live comfortably in most ordinary situations; should we get overly excited, and even contemplate changing our *logic*, because of the "special case" represented by sea anemones?

Labeling something a special case, or an "exception," is a strategy, if a simple-minded one, for avoiding to come to terms with what that case is screaming to bring up. This is the important truth conveyed by the saying "The exception proves the rule": if we are ready to recognize some violation of a rule as an exception, rather than evidence that the rule is wrong, this shows what powerful hold the rule has on us (so the exception "proves" the rule logically, not empirically: that something be assigned the *meaning* of an exception confirms that what it is taken to be an exception to is assigned the meaning of a rule). But here the strategy will not work, for the exception has a perplexing tendency to reveal itself as a rule.

It's not just a small matter of sea anemones: of those creatures that are themselves marginal. *All* the predicates we use in ordinary language, including the ones most central (anything other than marginal) to our concerns, have margins and hence cannot be neatly encapsulated in their boxes. (Marginal creatures like sea anemones are thus pointers to the destiny of all creatures.) In fact, the margins themselves have margins, of indeterminate size—a point that will become relevant shortly. All the predicates we use are *fuzzy*, or *vague*: indeterminate to an extent. Eubulides of Miletus, a philosopher of the fourth century B.C., a member of the Megarian school and a powerful intellectual adversary of Aristotle (the same guy, by the way, who first thought of the Liar), made this apparent with his celebrated paradoxes of the Bald Man and the Heap: A man with no hair at all is bald; if you add a single hair to a bald man, he remains bald; hence all men are bald. A single grain of sand does not make a heap; if you add a single grain of sand to something that is not a heap, this addition will not make it a heap; hence no heaps exist. There is an indefinite margin between heaps and non-heaps, and between bald and non-bald men, that threatens to swallow the difference between the two; and, what

---

[13] "The sea-anemones or sea-nettles, as they are variously called, are not Testacea at all, but lie outside the recognized groups. Their constitution approximates them on one side to plants, on the other to animals. For seeing that some of them can detach themselves and can fasten upon their food, and that they are sensible of objects which come in contact with them, they must be considered to have an animal nature. The like conclusion follows from their using the asperity of their bodies as a protection against their enemies. But, on the other hand, they are closely allied to plants, firstly by the imperfection of their structure, secondly by their being able to attach themselves to the rocks, which they do with great rapidity, and lastly by their having no visible residuum notwithstanding that they possess a mouth" (*Parts of Animals* 681a–b).

counts most, a similar margin exists between any two neighboring qualities—between *orange* and *red*, between *war* and *peace*, between *male* and *female*. You can define the two terms of any such opposition all you want; you still must face the fact that often your definitions will not be able to tell you which is which. The Heap and the Bald Man can be generalized to all ordinary predicates, giving rise to a perverse, puzzling kind of argument that is called in common parlance a *slippery slope* and in philosophical jargon a *sorites*.

The mystery of vague predicates has haunted contemporary formal (analytic) logic for decades and, despite the valiant efforts made to resolve it and the great talent displayed in such efforts, no progress has been made. At one point, it looked like *supervaluations* might do the trick: for each predicate, you have clear cases of it, for which all valuations in the relevant class agree on a value *true* and the supervaluation on them records that agreement; clear countercases, for which all relevant valuations agree on a value *false* and so does the supervaluation on them; and cases that are not clear at all, where the relevant valuations disagree and the supervaluation on them records no value. There are clear cases of things that are heaps (or bald, orange, or male), clear cases of things that are not heaps, and cases of things for which it is hard to tell whether they are heaps or not; and we should leave the heap-ness of those last ones indeterminate. But then it was pointed out that (as I indicated above) there is a *meta*vagueness, a vagueness of vagueness: margins have themselves margins; the border between the clear cases and the ones that are not clear is itself not clear, and the same holds on the other side; so that strategy collapsed. Others tried to make do with a *fuzzy logic* that allowed for infinitely many truth-values instead of the usual two; but, as Mark Sainsbury (1996, p. 255) remarked, "you do not improve a bad idea by iterating it"—there is going to be exactly the same difficulty negotiating the border between any of the infinitely many neighboring truth-values thus allowed as there was with the original two.[14] And there were even those who, by distinguishing metaphysics from epistemology, effectively threw up their hands in despair claiming that a sharp distinction between *bald* and *non-bald* does exist, but we don't know what it is. So the anomaly (for that is what it is) is here (that is: within the range of analytic logic) to stay. We can *artificially* (or, once again, *technically*) set exact boundaries for certain predicates to apply: we can decide, for example, that it takes a certain exact SAT score to be admitted to a certain college, a certain exact number of days to be considered for US citizenship, or a certain exact number of misdemeanors for the next one to turn into a felony; but that decision will always be felt by some (I suggested in Chap. 1) as an arbitrary imposition (as dependent on someone's *arbitrium* or act of will—which happens with all technical definitions), and in all other cases, failing similar arbitrary moves, the borders between predicates will remain hazy and the firmness promised by analytic definitions will be a mirage. It will be sea anemones all the way down.

---

[14] In Chap. 9 we will see that this strategy, and the problem just mentioned about it, mirror the strategy of representing continuous structures like a geometrical line by the use of infinitely many numbers, and the problem that other strategy runs afoul of.

Numbers are not in space and time and, though geometrical figures (which are also out of time) have spatiality, we do not encounter them in the space we inhabit: they are just as abstract (i.e., separate) from the concrete, spatiotemporal landscape where we conduct our ordinary dealings as numbers are (spatiality itself is for them an abstract quality). So are Platonic forms, concepts, essences, and all those other beings that I listed above and to which analytic logic applies in the most obvious, direct, natural way. For good reason: what is in space and time goes through *changes*, from being determined in certain ways to being determined in other ways; hence it is likely to contradict the irrevocable determinations asserted by that logic. In this area, then, we are going to find our third major anomaly: among *spacetime continuants*.

Take Socrates at five years of age. He has a number of qualities, and you can use some of them to tell him apart from any other child of the same age and from any adult; you can, based on those qualities, give a definition of him. Then take Socrates again at the age of 70, perhaps as he is participating in his trial. All the qualities he had as a child, and that as a child distinguished him from anyone else, are gone by now; most of the qualities he has now are even the opposites of qualities he had back then; the definition you provided of Socrates all those years ago no longer applies; and yet he is the same person—he is *identical* with the five-year-old he once was. How can that be? How can someone, or something, have so little in common with what he, or it, was and still maintain his, or its, identity? How can the ship of Theseus, after every plank of wood that constituted it has been replaced by a new plank, still be the ship of Theseus? What if (to borrow from Hobbes[15]) there were another ship from which all the new planks came, and there was not a simple replacement but an exchange made between the two ships, and in the end either was constituted exactly of the planks that once constituted the other, and still we thought that both had never for a moment lost their identity and their distinctness from each other?

Aristotle faces this issue in Book VII of the *Metaphysics* but does not go too far with it. His general approach is to claim that there is an *essence* of Socrates that remains unchanged for as long as Socrates exists, and that is constituted of all the qualities that are, indeed, *essential* to him; the collection of those qualities would then determine the identity of Socrates, *what it is* to be Socrates. But the only qualities essential to Socrates, in the Aristotelian scheme of things, would seem to be *animal* and *rational*, and aside from the fact that most of us would still recognize Socrates (or, say, our grandfather) as himself if he lost his mind (became a victim of Alzheimer's; here the neat divisions favored by analytic logic show their painful

---

[15] "[T]wo bodies existing both at once would be one and the same numerical body. For if, for example, that ship of Theseus, concerning the difference whereof made by continued reparation in taking out the old planks and putting in new, the sophisters of Athens were wont to dispute, were, after all the planks were changed, the same numerical ship it was at the beginning; and if some man had kept the old planks as they were taken out, and by putting them afterwards together in the same order, had again made a ship of them, this, without doubt, had also been the same numerical ship with that which was at the beginning; and so there would have been two ships numerically the same, which is absurd" (1839, pp. 136–137).

edge), those are the very same qualities that would constitute the essence of Callias (or of our grandfather), so we end up with the following inconsistent tetrad:

(21) Socrates is identical with the essence of Socrates.
(22) Callias is identical with the essence of Callias.
(23) The essence of Socrates is identical with the essence of Callias.
(24) Socrates is not identical with Callias.

And you know how badly contradictions fare in Aristotelian, analytic logic.

Once more, valiant efforts have been made to understand this as a problem and to resolve it. Some thought of introducing *individual essences* (occasionally also called *haecceities*), constituted of the same qualities but distinct *solo numero*—which (despite, again, my personal sympathies for this line of thought, made apparent in my 2013) seems to be an ad hoc manner of addressing obscurum per obscurius: the status of these entities is not explained, other than by saying that admitting them would resolve the problem. Locke proposed that memory, not essence, be a criterion of personal identity; but Kant pointed out, with a simple mental experiment, that the functional continuity of memory did nothing to justify substantial (metaphysical) continuity.[16] Others split up spacetime continuants into (instantaneous) *time-slices*, which was all well and good as far as the splitting went (analytic logic, as is to be expected, has its easiest time analyzing things) but became a nightmare when one had to explain why those particular time-slices, and no others, would have to come together so as to establish precisely what we intuitively understand as the boundaries of the life of a particular spacetime continuant.

None of this is going to persuade anyone—certainly not in the definitive way demanded by analytic logic. The commitment to this logic is the commitment to a paradigm, in the Kuhnian sense; those committed to it will continue to fault themselves, not the theory, for what trouble they run into. All that can be done, after accumulating large evidence of trouble, is bring out alternative paradigms and show that the trouble is easily dispelled there. Fortunately, and in contrast with Kuhn's (1962) examples (but consistently with those of Giambattista Vico 1984), we do not have to wait for such alternatives to succeed each other in time: in our case (and, I believe, most others) they coexist as distinct, systematically organized points of view on the same phenomena. So it is time to consider one systematic alternative to analytic logic.

---

[16] "An elastic ball that strikes another one in a straight line communicates to the latter its whole motion, hence its whole state (if one looks only at their positions in space). Now assuming substances, on the analogy with such bodies, in which representations, together with consciousness of them, flow from one to another, a whole series of these substances may be thought, of which the first would communicate its state, together with its consciousness, to the second, which would communicate its state, together with that of the previous substance, to a third substance, and this in turn would share the states of all previous ones, together with their consciousness and its own. The last substance would thus be conscious of all the states of all the previously altered substances as its own states, because these states would have been carried over to it, together with the consciousness of them; and in spite of this it would not have been the very same person in all these states" (1998, A363–364fn).

Interlude: A Conversation Between Two Analytic Neighbors

Jack: Hi Don. How are you doing?
Don: Just the person I was looking for. I need to talk to you.
Jack: What about?
Don: It's about our backyards. See, it's been a while since we moved in...
Jack: A couple of months.
Don: Indeed, and so far, we have been leaving everything open. Anyone can just wander back and forth between your property and mine.
Jack: Is there anything wrong with that?
Don: Yes, I think a lot is wrong with that. We should build a fence and share the expense.
Jack: That's some news. Why would I want to do that? I like the open space. I like the view. Why would I want to shut myself up in a cage?
Don: Well, to begin with, your dog always roams freely around here.
Jack: Immanuel is very friendly; he has never hurt anyone.
Don: Hasn't hurt me or my family. But suppose we want to have company, a barbecue for a few friends, a party.
Jack: He wouldn't bother them either.
Don: Maybe not, but some people just don't like dogs. They are afraid, and dogs can feel that and become aggressive.
Jack: Immanuel would never react that way.
Don: Still, I would not want my guests to be troubled.
Jack: Even if you knew that there would be nothing to worry about?
Don: Yes, because *they* wouldn't know it, or they wouldn't believe it, and that's enough. If they don't want your dog here, your dog shouldn't be here, end of story. There should be a fence that keeps him inside your property.
Jack: Just tell me when you're having company and I'll keep Immanuel inside the house. I don't see why we should pay for a major construction project because of something that will happen only once or twice a month.
Don: Well, it's not just that. There's also the issue of privacy.
Jack: What do you mean?
Don: In my own home, and in my backyard I should feel like I'm in my own home, I should be able to do whatever I please. Go around naked, if I want to, or make love in the grass. I shouldn't constantly have the impression that I am being watched by my neighbor.
Jack: I don't watch you, Don. I couldn't care less what you do.
Don: It doesn't matter if you watch me or not. What I said is that I should not *have the impression* that you are watching. And with all this open view, I have that impression. I don't feel comfortable in my own home.
Jack: Don, what you are saying is ridiculous. See how our properties slope on this hill? See those other houses further up, no more than a hundred feet away? Even if you raise a fence to protect yourself from my eyes, what's to prevent those other people from watching? Are you going to build a fortress all

Interlude: A Conversation Between Two Analytic Neighbors

around you? For all you know, anyone of them could be directing a telescope into your backyard and observing every detail of what happens there, even recording it for posterity if they wish, and there is nothing you could do about it because they would be on *their* property, doing what makes *them* feel at home.

Don: If they did that, I'd sue their asses. But why don't you let *me* worry about that and take care of it as I see fit, if and when I have a problem with it? I am trying to resolve an issue I have with you, and it's irrelevant whether or not I have issues with anyone else.

Jack: Alright, alright. So what about this? When you want to go around naked or make love in the grass, you just send me a text that I should stay indoors, and I'll oblige.

Don: You are unreasonable! You prefer to shut yourself up and your dog, just so that you can keep this space open. Why do you care so much about it?

Jack: People are different, we have to accept that. I enjoy the openness, the view, the sense of freedom it gives me. And I know that you will not always go around naked or make love in the grass; so most of the time I can be out here and even get to see you and have a nice conversation with you—not about a fence I hope.

Don: And because you want to enjoy the view and your sense of freedom, I should be forced to send you a text every time I want you out of the way? What about *my* freedom? The freedom to forget that you even exist?

Jack: There are pros and cons to every situation, Don. I am asking you to stand the little inconvenience of letting me know when you don't want me here, but at the same time if you do want me, if you need any help, you can just call on me and I will come to assist you, without the trouble of ringing bells, getting keys, or opening doors.

Don: Actually all this easy coming and going you seem to value so much is another reason why we should have a fence.

Jack: Oh no... There are other reasons?

Don: Sure. Did you hear about the break-in last week?

Jack: No, I didn't. What happened?

Don: A ne'er-do-well came into the neighborhood looking for something to steal.

Jack: To buy drugs with, I would think.

Don: Most likely. A guy at the bottom of the hill left his house key under the mat for his cleaning lady, and the crook found it and entered the house to steal jewels and computers and other nice stuff. He even took a briefcase to put everything in.

Jack: Very thoughtful of him. But why are you telling me this story now? What does it have to do with us and with your fence?

Don: It has lots to do with it. The thug entered that guy's house but *only* that house, because the house was fenced in on all sides. If the situation had been like ours he would have robbed us both.

Jack: Don, I don't have a cleaning lady, you know it. And I never leave my key under the mat, or in a pot, or anywhere outside the house.

Don: That may very well be. But how do I know that you won't forget to lock your door sometime or that someone else in your family won't?

Jack: Do you think we are stupid or senile? Do you think we don't care about being robbed, without you doing the worrying for us?

Don: You are probably the most responsible people in the world, but why should I base my own sense of safety on trusting you? Why should I make myself dependent on that trust? Every point of entrance into a house is also a weakness for it, and I have enough to worry about when it comes to the weaknesses of my own house without having to fret about the fact that every point of entrance into your house is also a point of entrance into mine.

Jack: Look, Don. You called me unreasonable, but it's what you say that doesn't make any sense. Because there may be times, *rare* times I take it, when you want to have a barbecue in the backyard, or even rarer times when you want to go around naked or make love in the grass, or because it is remotely possible, though highly unlikely, that sometimes one of us will leave our door open and that on just that occasion a crook will be prowling the neighborhood and be so lucky to find it open—just because of all these small contingencies that will never amount to more than (and I am being generous) 10 days out of the year, I should wake up all 365 days of the year to contemplate the wall of a cage! Don't you see what an imposition you want to enforce on me?

Don: You don't understand. It doesn't matter how many days I actually do those things. *I* feel caged, constrained by your presence all the time, all 365 days; *I* feel that it is an imposition on me to have to inform you when I want some privacy. I told you: I don't want to be forced to think about you all the time. If I care to see you and talk to you, I will come to your door and ring your bell; if I need your help, I will text you. But most of the time, I just want to be left alone, with you totally out of the picture.

Jack: As I said, Don, people are different. You clearly feel one way and I feel another, and our feelings are irreconcilable; we feel contrary things.

Don: But don't you see that my feelings should count more?

Jack: Now, that's great. Why would that be?

Don: Because the feelings that count the most are *hurt* feelings. Pain matters more than pleasure. Suppose someone takes pleasure in torturing another, and when the other guy tells him to stop he says: we are different; I would feel violated if you denied me this form of expression as much as you feel violated by the pain I inflict on you. Would that make any sense? Do you think there is any parity between these two contrary feelings? Should they be treated the same?

Jack: No, they should not. But who is the torturer here?

Don: You are. You are telling me that I should suffer my anxiety and my exasperation in silence so that you can have the pleasure of waking up to an open view.

Jack: But don't you see that your description could be reversed? Then it would be my pain of waking up in a cage that counts less, for you, than your pleasure of finally being relieved of my presence!

Don: Just like the torturer. He could reverse the description too and say that his pain in not being able to express himself should count more than the pleasure his victim would feel in being relieved of his presence.

Jack: But that would be absurd!

Don: No more absurd than your own redescription.

Jack: Wait a minute. The torturer is inflicting *physical* pain, is literally destroying the body of his victim. The pain he would feel if they made him stop, even admitting that we can allow such talk, would only be psychological; so clearly, there is no parity. But we are *both* talking about a psychological condition.

Don: Sure, Jack, but there are also *psychological* torturers. People who put others in double-bind situations, with no way out, no viable option for how to behave, and end up destroying them to an extent even more radical and irremediable than many physical torturers.

Jack: So I would be torturing you psychologically?

Don: That's what I have been saying.

Jack: And your proposal for how to remedy that is to torture *me* psychologically.

Don: I don't see it that way.

Jack: I know that you don't see it that way, but it is your word against mine. So all you are saying is that your pain should count more because your word counts more.

Don: I see that there is no way to resolve this.

Jack: No, there isn't. But I cannot prevent you from raising all the fences you want on your own property. I won't like it, just as I wouldn't if you planted a tree that blocked the view; but I can't decide for you. So go ahead if that's what you want; but be sure that you stay clear of everything that is mine. And be sure that I won't pay a penny for it.

Don: If I had been happy to do what you just said, I would not have needed to talk to you, and I could have avoided all this aggravation.

Jack: Yes, you could have, and I could have too. I could have relaxed down there in the shade instead of listening to your complaints. We have been wasting our time. Rather, *you* have been wasting *my* time.

Don: I don't think so. Now that we have agreed that there will be a fence, we must get to the sharing of the costs of it.

Jack: There will be no sharing, I told you. You do it on your own in your own backyard.

Don: I don't think that's fair.

Jack: Fair? You want to force me to do something I don't want and I don't like and want me to pay for it, and I am not being fair?

Don: Sure, because once the fence is up you will profit from it.

Jack: I will hate every inch of it!

Don: You say so, but it may be a tactic for having me take care of the whole project. Once it's up, you will enjoy privacy and security as much as I will, at my expense.

Jack: I told you I don't care about privacy and security—about *what you call* privacy and security.

Don: Yes, you told me. But the privacy and security the fence will provide you are objective goods, and for all I know, they might even increase the value of your house. That you say that they have no value does not make it so.

Jack: They have no value for me.

Don: Even if I give you the benefit of the doubt and believe that you not playing games with me, all I can grant you is that they have no value for you *now*. Tomorrow, or next month, you will see the light, or you will sell the house, *at a higher price*, and the new owner will see the light and appreciate the privacy and security the fence gives him, and I will have been paying for it.

Jack: OK, so suppose I agree that privacy and security are objective values. Do you agree that openness and freedom are also objective values? And that I have a right to set my own priorities among all these values? That I can decide to value freedom and openness *more* than your damned privacy and security?

Don: There must be an objective way of settling this dispute. There must be experts who can determine how to rank these values objectively.

Jack: Yes, and there must be laws against the kind of harassment you have been subjecting me to. There must be ways of protecting an honest citizen from this invasion of privacy—of *my* privacy, that you seem to have no respect for.

Don: I have a right to interrupt your lazy contemplation of openness to bring up an important matter.

Jack: I don't find this matter at all important. I find it petty, not to mention intrusive.

Don: That's another thing we see differently. But let me give it another shot. You have agreed that privacy and security are objective values, and I may agree that freedom and openness are too.

Jack: Now that's making progress!

Don: You also said that you rank freedom and openness higher, but perhaps you can tell me *how much* higher. Do you think that freedom and openness are worth twice as much as privacy and security, or three times?

Jack: A hundred times!

Don: Don't be foolish. Give me a reasonable figure.

Jack: Ten times.

Don: Then what if I asked you to pay for one tenth of the fence and to build one tenth of it on your property?

Jack: I would say, No.

Don: But why?

Jack: Because *you* should be compensating me for the loss of something that I value so much more!
Don: So no compromise is possible.
Jack: No. Build whatever cage you want, on your property, and leave me alone.
Don: You will be hearing from my lawyers.
Jack: *My* lawyers will.

Chapter 3
Dialectical Logic

Abstract In Hegel's dialectical logic, arguments never come to an end, and continuing them often results in proving consequences contrary to what was proved before: by continuing an argument that proves the mortality of Socrates, we may end up proving his immortality. Since not even contrariety can determine radical differences, those who reason according to this logic (like Hegel himself) have a tendency to deny such differences, and to think of world history as a single connected narrative—they have a tendency to *monism*. But note that, insofar as the word "narrative" is suggestive of a *temporal* development, this suggestion must be resisted: that one thing follows another in time must still be explained logically; history itself must turn from chronicle to demonstration. Time presents us with the immediacy of dialectical development; but this immediacy must be redeemed by being mediated conceptually; time parameters are promissory notes to be paid off by providing an account of why certain things did not just follow but *had to* follow certain other things—at which point the time parameters can be dispensed with.

Optimism is also naturally forthcoming in this framework, not so much because a point of view is available which is external to the narrative and argues for a positive resolution of it as rather because each phase of the narrative can only be spelling out its own (internal) values and seeing them reflected in the ways in which that phase is in fact turning out. On the other hand, dialectical logic cannot be formalized, because the whole context in which an argument is phrased (which, tendentially, is really the *whole* context) must always remain available for the next steps of the argument to be made. No abstraction from content is ever going to be legitimate.

The distinctive feature of analytic logic, I said, is not so much that meaning is obtained by a process of breaking down as rather that this process is irrevocable, for both words and sentences: what is obtained at the end of it is indeed an end—nothing further will be learned about a meaning, and certainly nothing that contravenes what the process has taught us. In Chap. 1 I described a confrontation in which one interlocutor took the meaning of a word ("liberty") to be evolving between the stage represented by his opponent and the one represented by himself; and in the third anomaly discussed in Chap. 2 I suggested that the "meaning" of a spacetime continuant (if that is the right word for it; more about this later) also evolves from one stage to another of that continuant's career. Both the confrontation and the anomaly point us in the direction I intend to take here; but it will help if, before I begin to travel in that direction, I offer a couple of examples that give an idea of where I am going. In my first example, I am going to argue that, contra the verdict reached above, Socrates is immortal—though remaining mortal as well. It goes like this:

> To be sure, Socrates died in 399 B.C. by drinking hemlock in prison, which makes him mortal; but that is hardly the end of the matter. For, who *is* Socrates? There was once an aggregate of cells that fulfilled various organic functions and was socially identified by the name "Socrates" as he walked through the streets and houses of Athens; but is that what we talk about now when we talk about Socrates? Not quite: we talk about a paradigm of morality that (much as a gadfly would with a large and sluggish horse) constantly stimulates every one of us, and warns us against making an exception of ourselves before the laws even when we suffered an injustice, and asks us never to reciprocate evil. This paradigm was once embodied in an aggregate of cells wandering around Athens, but has long transcended that terrestrial abode. It is now a perpetual icon, more alive than those cells ever were; hence it is immortal.

Superficially, the way this argument works seems to depend on a temporal development and, if we consider what goes on here to be another case of a meaning "evolving," we must note that evolution, in the most common understanding of it, is a process taking place in time. Eventually, I will have to dispel this appearance and claim that time has nothing to do with the current topic; in preparation for making that step, let me bring out here the important difference between time and temporality (analogous to the one noted above between space and spatiality). Time is, depending on one's metaphysics, a *real* structure within which the universe exists (a real container for it) or a *real* parameter of the universe's existence or of our experience of it; temporality is a structure or a parameter that is pertinent to all sorts of situations and circumstances—real or *unreal*. The adventures of Odysseus or Don Quixote have a temporal character; they occur one after the other; they last a few days or a few years; but such temporality has no place within (real) time; moreover, the temporality that is pertinent to Odysseus has no place within the one pertinent to Don Quixote, and vice versa—Odysseus and Don Quixote share no clocks. A distinction must be made between the single *chronological* time, in which we believe our lives to have their course, together with, in general, all the lives of things and people that *are*, that exist, and *narrative* time, which is the temporal dimension of a story: a seriality that implies no uniqueness and no being. History, of course, can be, and often is (perhaps even must be), told as a story, which will become relevant later; but even then the conceptual distinction still stands between its nature as a story and its nature as the telling of what in fact happened. The evolution of Socrates I referred to has a (somewhat dubious) claim to having in fact happened; but that is not what plays out in my argument, and similar arguments could be run about Odysseus and Don Quixote—about how *they* evolved, *in* the stories they belong to. Therefore, until further notice we will assume that the time pertinent to my argument and similar ones is narrative, not chronological. And, before that further notice, I will offer another example, this time proving that Plato, though remaining Greek, is also not Greek.

> To be sure, Plato was born in Athens and, since Athens is a Greek city, he is Greek. But does he, really, belong to Greece? The seed of Plato was sown in Greece; but then it grew into a large and beautiful tree, and now we need to ask ourselves, "What is *the truth of* that seed? What is the truth of Plato?" (In Chap. 5, this use of "truth" and its cognates will be thema-

tized; for the moment, we can understand it somewhat loosely—as if I was asking "What is the seed (or Plato) all about?") It is—we will answer—an articulate vision of humankind and society in which our biological ties, hence also our ethnic ones, no longer matter, and we must regard as brothers and sisters, and parents and children, all the people of certain ages who convene with us in a political community ruled by reason, in which everyone is assigned tasks based not on hereditary specifications but on how she can best contribute to the common good. So, at some point in the development that went from Plato the seed to Plato the tree, of that development which *is* Plato, Plato overcame his Greekness and became a patrimony of all humanity. When we say that he is Greek, we are referring to a phase of his being in which his truth was still germinating and had not yet come out in full force, hence it was not apparent to the general consciousness or even to him; in living out his destiny, he denied that primitive, though still genuine, qualification of himself. (Proof that Bruce Springsteen is black—while of course also remaining white (hence not black)—can be left as an exercise to the reader.)

Some of the terminology used in this example broaches details of the alternative logic I am introducing here and (I said) will become clearer as we go. For the moment, let me pause to observe a feature of the example that is not so obvious in the previous one: though I described what I did as a proof that Plato is not Greek, it is more directly a proof that he has no nationality, from which there follows that he is not Greek. It is, then, more directly (and consistently with previous remarks) a proof of a contrary, not of the contradictory, of Plato's Greekness. (For the previous example to have the same feature, I would have to turn it into a proof of the fact that, say, Socrates is a god, or maybe a daemon such as the one he occasionally invoked—in any case some contrary of *mortal* from which its contradictory *immortal* would follow.)

The above are two examples of the only alternative to Aristotelian logic that was made explicit in our tradition in painstaking and impressive detail: Hegel's *dialectical logic*.[1] As a first approximation to an account of this logic, the meaning of a word in it is not a collection of traits but a narrative, a story, in every phase of which the word is associated with a definite collection of traits but then those traits change, sometimes into their opposites, much as in ordinary stories characters go through crises and transformations, occasionally catastrophic ones, and the necessity that applies to the logic is narrative necessity: much as characters in a well-constructed (ordinary) story can go through crises and transformations while remaining recognizably themselves (the reader or spectator, that is, though sometimes startled by the uncharacteristic and unexpected turns of events, is able to recognize them as *the same* characters they were before and to acknowledge that this new version of them is what they would *have had to* become in the new, startling circumstances in which they are now to be found), so the various phases of the meaning of a word must be

---

[1] In my 2000 I claim that Hegel's main contribution to philosophy, whatever the ostensible subject matter he was at any time dealing with, was how he thought, reasoned, and argued about it, rather than any substantive thesis he might have had about it. Hegel himself did not always explicitly put it that way, and in any case reserved specific treatises to an account of his logic. The two major ones are the *Science of Logic* (1990) and the *The Encyclopaedia Logic* (1991a).

connected by so strong a narrative tissue as to convince a reader or spectator that a given phase of meaning is not just followed but *must* be followed by the next one.[2]

What are anomalies in one paradigm have a way of turning into natural cases of another (and vice versa); and the present pair of paradigms is no exception. Most obviously, spacetime continuants are role models for dialectical logic: they *show*, though they don't explain (Hegel would say: they provide the *immediacy* of it, not yet mediated by conceptual articulation—but destined eventually to receive such mediation), how there can be, indeed there is, identity in difference: how something can stay identical with itself while changing into something else—something that is even in some respects a contrary of what it was before. And it does not stop there. Within analytic logic, spacetime continuants are uneasy (though perplexingly frequent) exceptions to a general strategy of dividing and conquering: whenever a contrariety is found to infect something A, A must, for an Aristotelian, break down into two distinct things. If choice, for example, is something we do not share with other animals, and appetites we do share with them, then choice cannot possibly be an appetite: choices and appetites must (irrevocably) part their ways.[3] With spacetime continuants, on the other hand, the phenomena of ordinary discourse we are trying to explain force us to allow that, say, Socrates at 70 be 5 ft tall, Socrates at five not be 5 ft tall, and yet Socrates at five be identical with Socrates at 70—however intricate it might be to make sense of that. Then, of course, Socrates at 70 is executed, and we breathe more easily (our logic does): whatever contrariety there arises now between his (?) corpse and his previous vital configurations can be captured by our usual logical means, and we are (finally!) reconciled with the fact that the live person and the corpse are distinct entities. Within dialectical logic, on the other hand, the identity of Socrates at five and Socrates at 70 can be perfectly well understood—that is, it can be understood, in proper philosophical (Kant would say: transcendental[4]) terms, *how that identity is possible*, whereas understanding it empirically would require the availability of a biography of Socrates consisting of a thoroughly connected and convincing narrative, which we do not have. But, if that is the case, what is so special about death? Death is just one more transformation Socrates goes through, and the difference between the live Socrates and his (!)

---

[2] A point that will surface repeatedly in what follows but can already be stated here is the following: stories are told about what has already happened (even those that are told about what is now the future must be told from a position in which what is now the future is past, or at least present). Which accounts for dialectical logic's essentially retrospective stance and (as I explain in my 2000) provides the only plausible way of understanding Hegel's repeated claims of being at the end of history—which would otherwise sound ludicrous.

[3] "Those who say ... [choice] is appetite or anger or wish or a kind of opinion do not seem to be right. For choice is not common to irrational creatures as well but appetite and anger are.... But neither is it wish, though it seems near to it, for choice cannot relate to impossibles, and if any one said he chose them he would be thought silly; but there may be a wish even for impossibles, e.g. for immortality.... For this reason, too, it cannot be opinion; for opinion is thought to relate to all kinds of things, no less to eternal things and impossible things than to things in our own power; and it is distinguished by its falsity and truth, not by its badness or goodness, while choice is distinguished rather by these" (Aristotle, *Nicomachean Ethics* 1111b).

[4] For this understanding of Kant's key term "transcendental," see my (1987).

corpse is one more qualitative gap that could be bridged by an ingenious, resourceful story, thus denying that the gap be of ontological significance and establishing as much continuity between these two phases of the story as there is between any that belonged to Socrates' official biography.[5]

This discussion illustrates how dialectical logic comes to be inclined toward *monism*: the view that there is only one thing. It is, once again, a philosophical (or transcendental) inclination: any occurrence of a contrariety *can* in principle be resolved (and the contrariety overcome) by an appropriate narrative link, except that it *will* be resolved only when that link is at hand, and empirically we most often fall short of such an ambitious goal. The man Hegel trusted this inclination entirely, which explains two aspects of his work. First, he *claimed* that there is only one thing, *spirit*, whose story is the total, all-inclusive story: the one of which the story of anything (and anyone, including Socrates, Plato, you, and me) is but a chapter. Second, he devoted all his efforts to *telling* the total story, in what were often highly ingenious and resourceful ways. He could not succeed and didn't: his efforts could not help leaving his claim unvindicated—there was just too much to do. But these efforts, in the face of their inevitable and predictable failure, make it apparent that his monism was not an independent metaphysical view: it was (to use Kantian language) set on him as a task by his logic (hence it was metaphysical only in the sense in which "metaphysics" and "logic" are two names of the same thing). And, if you think that this amounts to dialectical logic setting unreasonable, absurd tasks, consider that analytic logic has a corresponding inclination to (and sets a corresponding task of) dividing forever and ultimately having (to use other Kantian terms, adapted from the second Antinomy[6]) everything made of nothing. Most philosophers who embrace it are too concerned with saving the phenomena to go all the way in trusting this inclination; they rather take the (enormous) trouble of maintaining some level of identity in the midst of universal disintegration. But occasionally one of them shows us the true nature of the game, and it is especially interesting when he is someone like Bertrand Russell, who also took metaphysics to coincide with logic. Then we can see the universe pulverized into instantaneous sense data, and any ordinary notion of identity reduced to as much of a delusion as talk of the present King of France is.[7]

---

[5] Philosophy has traditionally been associated with, even defined as, meditation upon death, or consolation of death, or in any case a focus on death. (It was only in the twentieth century that Hannah Arendt suggested a philosophy centered on birth.) So it might be instructive to notice in passing (and consistently with claims I have already made, and will continue to make) that there may be a logical angle to that obsession: death is where the radical divisions required by analytic logic apply most obviously. Conversely, one might regard dialectical logic as a logic of life (and indeed Hegel sees life as the immediacy of the idea—as the dialectical idea simply showing up).

[6] "For assume that composite substances do not consist of simple parts: then, if all composition is removed in thought, no composite part, and (since there are no simple parts) no simple part, thus nothing at all would be left over; consequently, no substance would be given" (1998, A434/B462).

[7] "We can now define the momentary common-sense 'thing,' as opposed to its momentary appearances. By the similarity of neighboring perspectives, many objects in the one can be correlated with objects in the other, namely with the similar objects. Given an object in one perspective, form

One other feature of Hegel's philosophical attitude is also a consequence of his logic: his optimism, his conviction that the dialectical development of spirit is inevitably going to bring out more complex, articulate, rich phases of it.[8] What about regress, some will ask, what about dark ages, what about sophisticated civilizations lost and never recovered, what about the *victims* of history? Those would be fair questions to ask, and implicitly fair objections to raise, if Hegel's optimism was based on an independent source of values—if, that is, he had some independent criteria for judging when something was more complex, more articulate, richer, etc., in a word, *better*, than something else and then claimed that, as a matter of fact, spirit proceeds invariably toward this independently identified better outcome. But that is not how it goes for him: his monism entails that all criteria of value are part of the process itself—that there is no outside point of view from which it would be possible or legitimate to judge that process. The only criteria applicable to any stage of development are generated by that very stage; so it is a necessary, transcendental condition that they validate it. And they are the only ones applicable because they are the only perspicuous ones, reflexive of the stage's reality and hence able to understand it. At any stage there may be people (Kantians, say) wanting to assess their time on the basis of abstract, unfashionable, even (in their view) eternal criteria; but they will be deluded, and their very delusion will be part of how their time is constituted, a part to be evaluated, like everything else, by the criteria intrinsic to their time and to be found (by those criteria) as providential as anything else. All that an intellectual can do, according to Hegel, is "comprehend his time in thoughts"[9]; if he refuses to do it and unreasonably sticks to "his own bright ideas,"[10] he just

---

the system of all the objects correlated with it in all the perspectives; that system may be identified with the momentary common-sense 'thing.' Thus an aspect of a 'thing' is a member of the system of aspects which *is* the 'thing' at that moment.... All the aspects of a thing are real, whereas the thing is a merely logical construction" (1914, p. 96).

[8] This conviction extends to many supporters of empirical (as opposed to logical) evolutionary views. Near the end of his (1996), for example, Charles Darwin says: "And as natural selection works solely by and for the good of each being, all corporeal and mental endowments will tend to progress towards perfection" (p. 395). The way the necessity of dialectical optimism is justified below could shed some light on such sanguine (and, as seen from the outside, indefensible) pronouncements.

[9] "This treatise, therefore, in so far as it deals with political science, shall be nothing other than an attempt *to comprehend and portray the state as an inherently rational entity*. As a philosophical composition, it must distance itself as far as possible from the obligation to construct a *state as it ought to be*; such instruction as it may contain cannot be aimed at instructing the state on how it ought to be, but rather at showing how the state, as the ethical universe, should be recognized.... To comprehend *what is* is the task of philosophy, for *what is* is reason. As far as the individual is concerned, each individual is in any case a *child of his time*; thus philosophy, too, is *its own time comprehended in thoughts*. It is just as foolish to imagine that any philosophy can transcend its contemporary world as that an individual can overleap his own time or leap over Rhodes. If his theory does indeed transcend his own time, if it builds itself a world *as it ought to be*, then it certainly has an existence, but only within his opinions—a pliant medium in which the imagination can construct anything it pleases" (1991b, pp. 21–22).

[10] "[T]he essential point to bear in mind throughout the whole investigation is that these two moments, 'concept' and 'object,' 'being-for-another' and 'being-in-itself,' both fall *within* that

stops being a sane intellectual while other saner ones continue to comprehend him and his insanity, among everything else in their time, in thoughts. So Hegel's monism and his optimism are of one piece, and people (genealogists, psychoanalysts, deconstructionists, etc.) who wanted to use dialectical logic without committing themselves to one of them had to reject both: limit themselves to examining the dialectic of a text, of a dream, or of a historical juncture so that they could also see, from the outside that they thus had left, how that dialectic might have negative outcomes. (I will return to this point in Chap. 10, within a more general discussion of the interpretation of texts.) Also, these remarks and quotes show that Hegel's optimism goes hand in hand with his refusal to look into the future, let alone provide utopian recipes for it (from which one might derive a critical, *negative* stance toward the present and past): with his insistence that logical/philosophical work can only reconstruct the necessity of events *after they have taken place*. Dialectical accounts are essentially retrospective ones:

> A further word on the subject of *issuing instructions* on how the world ought to be: philosophy, at any rate, always comes too late to perform this function. As the *thought* of the world, it appears only at a time when actuality has gone through its formative process and attained its completed state. This lesson of the concept is necessarily also apparent from history, namely that it is only when actuality has reached maturity that the ideal appears opposite the real and reconstructs this real world, which it has grasped in its substance, in the shape of an intellectual realm. When philosophy paints its grey in grey, a shape of life has grown old, and it cannot be rejuvenated, but only recognized, by the grey in grey of philosophy; the owl of Minerva begins its flight only with the onset of dusk" (1991b, p. 23).

I used the words "ingenious" and "resourceful" a couple of times already, and that was not coincidental: these words point to another fundamental feature of dialectical logic. In order to account for it, consider how a proof is constructed in analytic logic. It often takes a large amount of ingeniousness and resourcefulness to do so: one must be open to all the details of the objects under scrutiny and of their multifarious properties and relations, and cannot know in advance which details (if any) will deliver the spark of intuition that gets the proof going. But none of that belongs to the *logic* of the situation: it belongs, rather, to the heuristics by which one *discovers* this logic. Once one discovers it, *that is*, once one happens to focus on the specific details that will do the job, one must let go of everything else; and it is only then that the construction of the proof begins, and its logic is finally exposed. I assimilated an analytic logical argument to a postmortem; by holding on to that metaphor, I can now say that what is properly logical in an analytic proof is not the live process of discovery but the autopsy that is executed when that process is over. Everyone who over the centuries searched for the Holy Grail of a logic of discovery or tried to conjure it up by using such fanciful, opaque words as "abduction" was not looking for analytic logic and was not content with it.

---

knowledge which we are investigating. Consequently, we do not need to import criteria, or to make use of our own bright ideas and thoughts during the course of the inquiry; it is precisely when we leave these aside that we succeed in contemplating the matter in hand as it is *in and for itself*" (1977b, pp. 53–54; translation modified).

So ingeniousness and resourcefulness do not belong in analytic logic: what is valued in it is rather automatic, machinelike, thoughtless proceeding. It is quite the opposite with dialectical logic. Say that I just proved in it that Socrates is immortal, by an argument like the one I gave above. The tendency of an analytic logician, at this point, would be to distill the "logical core" of the situation by distinguishing it from confusing "grammatical" irrelevancies and then to export this core to other situations, in order to prove that, for example, Plato is immortal too. In dialectical logic, that tendency must be resisted: even after the proof, not just Socrates but also "Socrates" remain very much immortal, and all the relations they have with anything else remain very much in play. Maybe the next thing we will prove about Socrates, by using other or the same details, is that he is mortal after all, or that his immortality is a form of mortality; and certainly, if Plato is also to be proved immortal, it must be by some other route originating in the specific semantic neighborhood where "Plato" is located. An ingenious, resourceful exploitation of *everything* that pertains to Socrates and Plato is constitutive of a dialectical logical treatment of them.

One could insist that, in proving that Socrates is immortal, I still selected some specific features of Socrates and built the proof from them; hence I still identified a logical core that was pertinent to the proof. I am not going to deny that, but I am going to point out that the selected features only have the logical significance they have because of the context in which they are situated: the context of *all of* Socrates. The proof is, crucially, not over with the fragment I proffered; it is an infinite process that, by incorporating fragments like that as its own chapters, only tells the truth about Socrates, and simultaneously gives the meaning of "Socrates"—or, for that matter, of "immortal"—when considered in its entirety; hence what contribution even this fragment makes to the articulation of that meaning and that truth can only be appreciated if one never loses sight of the entire context—of everything about Socrates, including what was not selected this time. So the two elements of the analytic tendency mentioned above show their intimate connection as dialectical logic reverses both: it is because the dialectical proof never ends that I cannot make a summary representation of it, take that summary to bring out a substantive logical point, and apply the same point elsewhere (I will never know if how *immortal* is to be understood elsewhere matches the way it is understood here); it is because no logical core of this proof can be isolated and exported elsewhere that I must continue to pay close attention to the context where I am working now and see what else ingeniousness and resourcefulness can extract from it beyond what was already obtained.

This strict dependence on context makes it impossible to turn dialectical logic into a formal exercise. There have been recurring attempts at doing just that, often inspired by statements of Hegel to the effect that, say, he had performed a "negation of the negation."[11] But those statements are summaries of work done, sketchy reminders of a process; and here no summary, I said, can ever substitute for the work, and no sketch provides the "essence" of a process. You can say that what you

---

[11] See for example his (1991a), p. 151.

did was deny a previous negation of Socrates' immortality, or a previous negation of Plato's un-Greekness; but a simple relation between negation signs, here, has no logical content whatsoever. Everything hinges on *how* you proved the immortality of Socrates, and that is going to be by different logical moves than those made in proving the un-Greekness of Plato, that is, by different sorts of appeals to the meanings of the relevant terms, as they structure themselves in the relevant contexts; hence to say, at the end of a good day's work along these lines, that what you did in both cases was to negate negations would tell you *nothing* about the work you did. It would have the same value as Leibniz attributes to Descartes' "method": "Sume quod debes et operare ut debes, et habebis quod optas" ("Take what is needed; do as you ought; and you will get what you want," 1880, p. 329). Therefore, not surprisingly, formalizations of dialectical logic are useless and pathetic: if in reviewing Hilbert's formalization of geometry you feel that you are getting to what actually drives the reasoning of Euclid, or of Gauss, in reviewing these other efforts you get the impression of missing all that matters in Hegel's reasoning and of throwing away the baby with the bathwater. A necessary condition for a logic to be formal—that is, as I said earlier, for it to maintain its integrity as a theory of meaningful discourse while disregarding some (even most) of the content of that discourse—is that such content *can* be disregarded without logical loss, without losing track of precisely the logical points we are trying to make, and that condition is not satisfied by dialectical logic; therefore, quite simply, dialectical logic cannot be formalized.[12]

There is, however, a distinction to be made in dialectical logic, too, between the logic itself and the work whereby the logic is unearthed. The exploration of a particular logical context and the tracing of ingenious and resourceful ties between various aspects of it take place in time; and we have seen that those aspects, often, have themselves a temporal dimension. The immortality of Socrates and the un-Greekness of Plato have been described as emerging in the course of time. I have contrasted chronological and narrative time and claimed that only the latter is in question here; but narrative time still is *time* of sorts, so time would seem to have crucial significance for dialectical logic. That is not the case, as was suggested earlier, and we need to address this issue now.

Using terminology that has already been brought up, time presents us with the immediacy of dialectical development: in time, identity in difference simply appears unannounced, whatever your logic—and if you are an analytic logician, I said, you will have considerable trouble accounting for it (i.e., saving the phenomena). If you are a dialectical logician, on the other hand, the immediate evidence thus provided will reassure you that you are on the right track; but your job is only to begin, not to end, there. For what you need to do now is explain that identity, and resolve it in the connectedness of a discursive path that will make you able to justify it, not just accept it at face value—that will make you *understand* what allows the identity to be maintained across the difference. The more you understand it, and the more connected your path becomes as every individual difference is smoothed over by it, the less relevant it will be that certain things have happened *at a certain time* or *tempo-*

---

[12] More detail about this issue can be found in my (2015).

rally after certain other things. Time parameters will be seen more and more as promissory notes for an explanation still to be completed, for a gap in your discourse that still needs to be replenished, as the whole context is gradually exposed and the relations that are operative in it do the actual work of articulating the discourse and exonerating the rough, primitive signposts by which those parameters marked the terrain. It will be hard work, and ultimately (as I suggested) unrewarding (it will never be done); but we know that at the (unreachable) limit of it the sheer specification of quantity contained in the description of Socrates as five years old would be transcended in a detailed text that makes that phase of Socrates necessarily grow out of any previous one, when he was a day (or an hour, or a second) less than five years old, and where as a result all reference to time positions, time measurements, and time occurrences could drop out.

So "meaning" is indeed the right word for what dialectical logic intends to find out about spacetime continuants: we know from the section on sense certainty, early in the *Phenomenology*, that Hegel regarded the demonstrative reference usually assigned to pronouns like "this" as just a confused case of conceptual reference,[13] and he would say the same of the one assigned to proper names. Not surprisingly, in view of this attitude, as well as of the practical unfeasibility of connecting all the dots of an individual's biography, the tireless effort he put out in reaching for the total story was mostly devoted to what the (analytic) tradition would consider universals.[14] For it is much easier—to the point that you might eventually (though misleadingly) judge it done—to prove how stoicism must develop into skepticism,

---

[13] "It is as a universal too that we *utter* what the sensuous [content] is. What we say is: 'This,' i.e. the *universal* This; or, 'it is,' i.e. *Being in general*. Of course, we do not *envisage* the universal This or Being in general, but we *utter* the universal; in other words, we do not strictly say what in this sense-certainty we *mean* to say. But language, as we see, is the more truthful; in it, we ourselves directly refute what we *mean* to say, and since the universal is the true [content] of sense-certainty and language expresses this true [content] alone, it is just not possible for us ever to say, or express in words, a sensuous being that we *mean*" (1977b, p. 60).

[14] The only exceptions are "world-historical individuals" like Napoleon. In a letter to his friend Friedrich Immanuel Niethammer on October 13, 1806, Hegel writes: "the emperor—this world soul—I saw riding out through the city reconnoitering;—it is indeed a wonderful sensation to see such an individual, who, concentrated here at a single point, astride a horse, reaches out over the world and dominates it" (1984, p. 114). And in (1991c) he explains: "World-historical men—the Heroes of an epoch—must, therefore, be recognized as its clear-sighted ones; *their* deeds, *their* words are the best of that time. Great men have formed purposes to satisfy themselves, not others. Whatever prudent designs and counsels they might have learned from others, would be the more limited and inconsistent features in their career; for it was they who best understood affairs; from whom *others* learned, and approved, or at least acquiesced in—their policy. For that Spirit which had taken this fresh step in history is the inmost soul of all individuals; but in a state of unconsciousness which the great men in question aroused. Their fellows, therefore, follow these soul-leaders; for they feel the irresistible power of their own inner Spirit thus embodied. If we go on to cast a look at the fate of these World-Historical persons, whose vocation it was to be the agents of the World-Spirit—we shall find it to have been no happy one. They attained no calm enjoyment; their whole life was labor and trouble; their whole nature was nought else but their master-passion. When their object is attained they fall off like empty hulls from the kernel. They die early, like Alexander; they are murdered, like Caesar; transported to St. Helena, like Napoleon" (pp. 30–31).

or mechanism into chemism, or art into religion. But we should always remind ourselves of what the task is: that *all* chronology be redeemed by what I called narration; that we be able to tell all of history as a good, connected story; and if that were to happen, then the very terms "narration" and "story" would lose most of their significance, and "proof" or "demonstration" would sound more appropriate. For then what the time in which narrations and stories take place was opaquely, uncomprehendingly pointing to—seriality: that one thing follows another—will be finally comprehended; it will be comprehended why something *must* follow something else; sheer juxtaposition (immediate following) will give way to logical consequence (conceptually mediated following). And we could imagine all of it "happening" in an instant—which is to say: out of time altogether, in an eternity which, like the instant, is atemporal.

With all the qualifications I made, it continues to be the case that spacetime continuants, at which analytic logic seems able to do little more than address a puzzled stare, offer the most obvious evidence that something like dialectical logic must be taken seriously. We are to expect, conversely, that the natural cases of analytic logic be anomalies for its dialectical counterpart; and such is what happens, most clearly for mathematics—the way mathematics is ordinarily understood. If you want, you can tell the meaning of "2" as a story; but it will be a tedious story, in which 2 and 2 always make 4; and if you were to discover new properties of 2, you would have to admit that they are new only in a manner of speaking, because, despite your ignorance of them, they had been there all along—the "story" is only one of you learning about them. There is more to mathematics than these stock examples and the ideology that goes with them, and we will get to it later (I will return to this issue in Chap. 9, where we will be able to see it differently); for the moment, I notice that it is just as frustrating (and for the same reasons) to apply dialectical logic to everything else that receives a natural analytic treatment—technical terms, concepts, essences, and the like—insofar as they are Platonically regarded as sitting in an immutable heaven. In all such cases, forcing a narrative format onto meaning will feel strained, artificial, and gratuitous.

Turning now to the other situations that constitute anomalies for analytic logic, dialectical logic will have a natural way of dealing with ambiguities. Every one of them will be perceived as a challenge to provide the connecting links that are missing, and that make ambiguities sometimes mysterious. It is no wonder that suggestions for the ingenious and resourceful activity that awaits us here can, and often do, come from poetry, as would be the case if I replaced the analytic fallacy (18)–(20) of p. 13 with an epigram like

(1) I played a sharp note and cut myself

Poetry, after all, is what makes language move, away from its entrenched settlements, from its clichés; so poetry is as immediate a presentation of dialectic—of the liveliness that dialectic projects—in the verbal world as spacetime continuants are in the physical one. A poem is a pulsating organism—one that proclaims its vitality on the page, rebelling against the very inscriptions in which it is locked, affirming a meaning that supersedes them—and a spacetime continuant is a poem incarnate, a physical structure that, like a poem, keeps defying any attempt at reducing it to a formula, any "definitive" characterization of it (analytic definition as such).

Sorites are another matter. Hegel makes room for them and puts them to use; but he would have to. Every logic (another point I will take up later) conceives of itself as universal, hence also conceives of all anomalies as problems to be solved—the already-mentioned agony of analytic logic over vague predicates is good indication of how this pretence works in it. But there seems to be a radical contrast between a sorites and a dialectical process: if I move from *pink* to *red*, it will not be by accumulating reasons why anything pink would also have to be red, that is, turn into one of its contraries; it will be by accumulating reasons why everything is pink, and there is no red. And what makes for even more trouble is that all of that would seem to depend on where I started—on my accidental decision to start on the pink side. If I had started on the red side, I would have reached a conclusion that is the mirror image of the previous one: everything is red, and there is no pink. (Similarly, the Bald Man and the Heap can be turned on their heads to prove that there are no bald men or that everything is a heap.) This perfect symmetry, too, flies in the face of dialectic: one can, indeed one must, admit that within the whole context relevant to dialectic everything hangs together, hence there are symmetrical relations of dependence between its various elements; but one cannot expect that these symmetrical relations be substantiated by exactly the same moves. The slave (we will see) depends on the master in a different way than the master depends on the slave. And again, this would be a mere oddity if all we had was the nuisance it offered; it can be read as suggestive only when we come upon something valuable which it could suggest. That, I claim, is what happens: sorites are pointing the way to a third theory of the *logos*, and it is time for us to start traveling that way.

Interlude: A Conversation Between Two Dialectical Neighbors

Jack: Hi Don. How are you doing?
Don: Just the person I was looking for. I need to talk to you.
Jack: What about?
Don: It's about our backyards. See, it's been a while since we moved in…
Jack: A couple of months.
Don: Indeed, and so far we have been leaving everything open. Anyone can just wander back and forth between your property and mine.
Jack: Is there anything wrong with that?
Don: Yes, I think a lot is wrong with that. We should build a fence and share the expense.
Jack: That's some news. Why would I want to do that? I like the open space. I like the view. Why would I want to shut myself up in a cage?
Don: Actually, that would not be shutting yourself up. It would be liberating yourself.
Jack: What do you mean?

Don: Think of when we were hunter-gatherers, hundreds of thousands of years ago. We went around in small groups and there was nothing impeding our march. No walls, no barriers, no fences. Except for natural obstacles, we were entirely free to roam the territory in any direction we pleased.

Jack: You are starting your own march from way back in the past. What do prehistoric humans have to do with us?

Don: A lot, as you will see. So these individuals were free to go and to stay wherever they wanted. They were also free to be attacked by other groups of hunter-gatherers and slaughtered, perhaps at night, when they were asleep, when they were helpless prey. They were free to eat what they caught, and free to starve when they caught nothing.

Jack: Just as everyone today is free to buy a ten-million-dollar mansion, though most don't have the money for it.

Don: Precisely: it's the same issue. This notion of freedom is too primitive, in a literal sense. There is nothing wrong with it, at the stage of the hunter-gatherers; but then people start thinking about it and they realize that they are, for all intents and purposes, drastically caged in. Not by walls or fences but by their precarious situation, by the deadly risks they constantly encounter, by the need to constantly defend themselves, by their constant fear.

Jack: A psychological cage.

Don: Yes, and a more effective one than anything physical. For you can break the bars of a physical cage, but from a psychological one you will never escape. It will always be blocking you, because it is *inside* you.

Jack: I am not sure that physical bars can be broken down so easily. But keep going; I am curious to see where this takes you.

Don: It takes me to a natural evolution of freedom. It turns out that people are more free, not less, when they restrict their original capacity to wander aimlessly and accept limitations in the spaces that belong to them, when they restrict their original capacity to make, at every moment, an arbitrary choice and subject their choices to agreements with their peers. Their freedom can best express itself only by incorporating these limitations, because now there are more things they in fact *can* do—including doing whatever they do in a safer environment, where it can be done better.

Jack: I sort of figured that you would get there. Now you are going to tell me that civilized humans have a more mature sort of freedom because they are integrated in a system of laws and a neat division of property that allows them to use what is uncontroversially their own in developing their own tastes and talents.

Don: That's what I would tell you if you didn't already tell it yourself.

Jack: And, presumably, you are also going to tell me that each of us is going to be freer if we just as neatly divide our properties. That my primitive sense of openness must evolve in an openness to be found beyond the building of a fence, not before it.

Don: That's right. Why do you think people say that good fences make good neighbors? Because all enjoy their homes more when their spaces are identified clearly; because it is easier to have healthy relations across a fence, free to choose when to make contact and when not to. Because that is what elementary, indefinite openness must graduate into: clarity. Just as a piece of writing is more open, it communicates more, when it abides by all the rules of spelling, and grammar, and syntax. To begin with, you may find those rules constraining—you certainly did as a child, when you were forced to learn them—but that very constraint is what ultimately frees your prose, frees you to write what you want to write, and to be understood by others. Even to be better understood by yourself.

Jack: Everything you said makes a lot of sense, so much that, as you have seen, I could say most of it myself, once I got the gist of it. I can do that because I have been there, because it is a path I have traveled, and in the houses I had before I was the one who went out and made the same speeches to my neighbors.

Don: Then we agree?

Jack: Yes and no. I agree that you are on the right track, but I believe you are stopping too soon.

Don: Now it's you who has to tell me what you mean.

Jack: I'll try. What you have done is show that things can turn into their opposites while still remaining themselves, indeed a better, more advanced form of themselves. A man who was short and weak as a child may turn into an adult that is no longer short and weak. A freedom that amounts to not knowing any rules or restrictions may turn into one that has accepted rules and restrictions, and that is *more* freedom because of that.

Don: Yes; this is what I have done. What's wrong with it?

Jack: Nothing is wrong. It is perfectly right, as far as it goes. Except that it doesn't quite go far enough. For the adult who is no longer short and weak is about ready to turn into an old man who is once more short and weak, though in a different way.

Don: And freedom?

Jack: Same thing, though it takes some effort, some ingenuity to see it. When you see it, on the other hand, it strikes you as the most natural thing in the world, the one you should have seen all along.

Don: Well, make me see it, then.

Jack: When you told your story, you ran together two distinct sorts of constraints: physical ones like fences (that supposedly make good neighbors) and legal ones like rules. So let me pick up the story where you left it off, and take it one or two stages further, so that you can see what I see now. In the old days of civilized humans, long after the hunter-gatherer era, physical barriers were predominant. People locked themselves in their fortresses, worried about venturing into alien territory, armed themselves to the teeth when they had to.

Don: Some still do.

Jack: Yes, and why? Because they don't trust the institutions, so they revert to the past. But they cannot live in the past; they must get on with the present, with the new ways things are turning out.
Don: And if they don't?
Jack: They will be left behind. They will be the pitiful leftovers of a bygone time.
Don: I don't know that they care.
Jack: That's *why* they are pitiful leftovers. But forget about them. My point is that eventually the constraints, the "cages" that proved more liberating were found to be the legal ones, since they allowed a new form of wandering around, at the same time unimpeded and secure.
Don: What are you talking about?
Jack: I am talking about all the public spaces that have opened up as civilization progressed. I am talking about parks, roads, squares, sidewalks. All places where people are restrained not by physical barriers but by the respect for each other that the laws require. So that they can go and stay wherever they please, like those cavemen you were talking about, without being afraid of others doing violence to them.
Don: Sometimes they do.
Jack: Sure: a growth process is not painless, and is not victimless. But it is also unstoppable.
Don: You are quite the optimist.
Jack: It's not my personal sentiments that matter. It is the logic of evolution itself, as much of social evolution as of the biological variety.
Don: I'm fine with everything you said; but I still don't see how it relates to our situation. We are not talking about public spaces here; we are talking about private ones, about my property and yours.
Jack: That's what I thought until a few years ago, and that's why I did have fences in those other houses. But then I realized that I had to go one step further. Public spaces are constituted by having everyone agree to leave them public, so there is no reason why two people like you and me could not agree to leave part of our spaces public, thus obtaining the same combination of liberty and security that people have on streets and sidewalks.
Don: I guess that pretty soon you'd want to make the insides of our houses into public spaces too.
Jack: Why not? That might be the next step. But progress is made one step at a time, so for the moment let's just enjoy the new phase of freedom our open backyards give us.
Don: I don't enjoy any of it. I don't enjoy your dog getting on this side and scaring my guests; I don't enjoy you watching me whatever I am doing; I don't enjoy the fact that someone might rob your house and then, finding no barrier between us, rob mine as well.
Jack: That's exactly as it should be. You should not enjoy any of that. Didn't I tell you that there are going to be growing pains?
Don: Who decides that these are *growing* pains? Why couldn't it be instead that *you* have been left behind by history, that you are a pitiful leftover over-

whelmed by nostalgia, craving for a primitive sort of freedom? Why shouldn't *I* push you along, make you get on with the present, with how things are turning out?

Jack: So you didn't find my explanation irresistible, unobjectionable?

Don: It's irresistible and unobjectionable from where you are sitting. From my point of view, it is just as irresistible and unobjectionable to claim that you should yourself take one or two steps forward.

Jack: But my account incorporates your point of view, just as your freedom incorporated limitations. So, in the same way, my account is a more mature development of yours, more up to date.

Don: Jack, this is a game that two can play. I can insist that your alleged "more mature" development is really only a reiteration of an obsolete stance, which takes itself to be more mature simply because it does not know any better: everyone can do nothing other than play out its own perspective, and what values are dominant in it. Or I can continue your story, add a new chapter: the one in which freedom goes back to requiring physical constraints.

Jack: I would love to hear that.

Don: You have been describing your position as the most novel, so novel that you are one of the few who arrived at it—one of the charismatic, historical individuals who point the way of the future. But to me that same position sounds outdated, a remnant from an earlier age. In the sixties and seventies of the last century, people had already gotten there: they lived in communes, they worked in offices without walls, *and* they thought that this would make them freer and merrier. When it didn't, when it started creating all sorts of problems, they understood something important and made another step: they realized that public spaces are liberating because we can decide when to enter them, and we can alternate between them and private ones. We can leave our private offices to go into a public area; we can leave our private houses to see friends in a coffee shop. We don't have to *live* in a coffee shop; we don't have to stand the presence of others if we don't want to. Privacy is what confers value on publicity. The new phase of freedom is, so to speak, freedom-square, *meta*freedom: the freedom of choosing between private freedom and public freedom.

Jack: That's ingenious. But don't you think that I could add a new chapter too?

Don: I'm sure you could. But I'm also sure you see that we could go on like this forever and the fence would never get built.

Jack: It would be built in a chapter and taken down in the next.

Don: And it would cost ten times more than just building it once and being done with it.

Jack: What is money if not the public circulation of goods?

Don: So do you want to make your money public, to share it with me?

Jack: Why not? Wouldn't you find that liberating?

Don: Not at the stage at which I currently am, not if that means that I have to share *my* money with you, or with anyone else that comes around. But I see

Interlude: A Conversation Between Two Dialectical Neighbors

	that you and I are out of sync, that our stages of development are. I should talk to you again when you have made some progress in your intellectual development.
Jack:	Or when *you* have.
Don:	This is hopeless, Jack. Perhaps I need to build my own fence in my own property, to express my own stage of development.
Jack:	Perhaps you need to do that, and then tear it down later.
Don:	Or wait until you will want to pay your share of it.
Jack:	Because I have finally reached your stage?
Don:	Yes, because you have grown out of your childish pretences.
Jack:	It's totally arbitrary for you to call *me* childish. That judgment is internal to your view, and totally superseded in mine.
Don:	OK, so I am going to go inside now and write the ultimate story, the one that incorporates everything that you have said, everything that you now think, every argument you could possibly use, and when you read it you will have to agree that that is the truth of it—the truth of what you yourself are now claiming.
Jack:	But not the truth *my* story will prove, the one *I* am going to write, incorporating that very agreement as an early chapter.

Chapter 4
Oceanic Logic

Abstract Hegel called his form of reasoning "logic" and, though many analytic logicians would disagree with his use of the term and judge it illegitimate, one cannot deny that there is an important tradition supporting it. The third logic to be discussed in this book, however, has never been recognized as such; so we must start by giving it a name. The name chosen here, though not entirely without qualms, is "oceanic." The significance of this logic is upheld by showing it at work in great philosophers of the past: Anselm, Spinoza, Bergson, Heidegger.

Oceanic logic understands opposite predicates and views as different ways of looking at the same thing, different modes of it in the way in which everything ordinarily taken to be an independent thing turns out to be, in Spinoza, a mode of God (or substance or nature). In it, sorites (which in analytic logic spell out the anomaly of vague predicates) are a valuable tool, as they reveal that whether something will be regarded as, say, white or black entirely depends on how one looks at it.

An ordinary practice whose ideal conditions seem to depend on the practitioners adopting an oceanic logic is *compromise*: whereas in analytic logic compromise can only be obtained by each party to a negotiation giving up something they want, and in dialectical logic it can only be obtained by a narrative step that incorporates and transcends all the conflicting views, in oceanic logic the views remain what they are, and since they are seen to be different views of the same thing, they all may end up obtaining exactly what they were looking for. Win-win is a natural outcome in it.

Analytic and dialectical logics are established disciplines, and just as established are their names. There is considerable philosophical work to be done legitimizing their common nature as logics, as well as accounting for their contrasting features; but we will be moving in areas that have already been mapped and where important vocabulary has been introduced. With our third logic, we are in uncharted territory, and I feel the way Aristotle must have felt when he recognized virtues no one had ever acknowledged before—say, the one pertaining to small honors or the one pertaining to anger. In some cases (like the first one mentioned) the Philosopher described the newly identified creature but left it nameless, probably because he did not regard it as sufficiently conspicuous; in others (like the second one mentioned) he gave it a name but was not entirely satisfied with it.[1] In my own case, I must give a name to what I am talking about, since I take it to be of equal dignity to what most people, including most philosophers, have been convinced for ages was logic, period, so I believe that, though my nameless subject matter might not be conspicuous at present, it *should* be (and it is regrettable that it is not yet); but I find it hard to hit upon a name that I am entirely satisfied with. In the past, I have called it *gradual* logic, and "fuzzy" would be another plausible candidate—except that it is already taken, to label something that I do *not* regard as a genuine alternative to analytic logic. Here I am going to try "oceanic," which makes sense, as we will see, but is a bit bombastic and grand. If my view catches on, I am sure that there will be more effective names proposed.

It would be a bad idea to try to motivate people to acknowledging a brand new theory of meaningful discourse based on the difficulty of resolving a paradox like the sorites—it would sound like sour grapes, or like using a cannon to shoot a fly. A more promising strategy is to bring out instead some *positive* examples of reasoning that (just as happens with the sorites) are not naturally understood in either analytic or dialectical terms—better still if they are associated with great philosophers, hence are suggestive of *good* reasoning. That is the course I am going to adopt now. My first example is a kind of prologue in heaven, borrowed from Anselm of Canterbury and his most famous work (if not his most imposing, for that would be the *Cur deus homo*): the *Proslogion*.

In Chap. 18, Anselm has already proved God's existence and all of his attributes—that is, his perfections—and has resolved a number of apparent conflicts among various perfections (between being merciful and just, say). But a new source of trouble surfaces and upsets him to no end:

> Once again, "behold confusion!"... Behold, once again mourning and sorrow stand in the way of one seeking joy and happiness. My soul hoped for satisfaction, and once again it is overwhelmed by need. I tried to eat my fill, but I hunger all the more. I strove to rise to the light of God, but I fell back down into my own darkness. Indeed, I did not merely fall into it; I find myself entangled in it. (p. 110)

What is going on? There is a problem: if God is shown to have many perfections, then maybe he is shown to be many *different* things, even to have different parts, which would itself amount to the lack of a specific perfection—of *simplicity*. For composite—that is, not simple—things can be destroyed, disintegrated: they can lose their integrity if their elements come apart. Such a fate, however, cannot possibly impend upon God; hence there must be something wrong going on here.

---

[1] "There seems to be in the sphere of honor also, as was said in our first remarks on the subject, an excellence which would appear to be related to pride as liberality is to magnificence.... The mean being without a name, the extremes seem to dispute for its place as though that were vacant. But where there is excess and defect, there is also an intermediate; now men desire honor both more than they should and less; therefore it is possible also to do as one should; at all events this is the state that is praised, being an unnamed mean in respect of honor" (*Nicomachean Ethics* 1125b). "Good temper is a mean with respect to anger; the middle state being unnamed, and the extremes without a name as well, we place good temper in the middle position, though it inclines toward the deficiency, which is without a name" (ibid.).

> Surely you are life, you are wisdom, you are truth, you are goodness, you are happiness, you are eternity, and you are every true good. These are many things; my narrow understanding cannot see so many things in one glance and delight in all of them at once. How then, Lord, are you all these things? Are they parts of you? Or rather, is not each of them all that you are? For whatever is composed of parts is not completely one. It is in some sense a plurality and not identical with itself, and it can be broken up in fact or at least in the understanding. But such characteristics are foreign to you, than whom nothing better can be thought. (p. 111)

The resolution (? see below) comes immediately (and was in fact intimated in the previous passage), as is always the case in this extremely compressed work. God's various perfections are not different things; specifically they are not different parts of God; they are always the same thing, phrased differently.

> Therefore there are no parts in you, Lord, and you are not a plurality. Instead, you are so much a unity, so much identical with yourself, that you are in no respect dissimilar to yourself. You are in fact unity itself; you cannot be divided by any understanding. Therefore, life and wisdom and the rest are not parts of you; they are all one. Each of them is all of what you are, and each is what the rest are. And since you have no parts, and neither does your eternity, which you yourself are, it follows that no part of you or of your eternity exists at a certain place or time. Instead, you exist as a whole in every place, and your eternity exists as a whole always. (ibid.)[2]

I am not saying that analytic logic would have a problem with this case. For all I know, it might approach it by stressing the distinction between a substance (here God) and its attributes (wisdom or goodness) and insisting that the substance can be one and indivisible though it has many attributes—that Anselm is just getting confused between attributes and parts of a substance. A slice is a part of a cake; the sweetness of the cake is an attribute of it; that the cake has distinct slices makes it a composite substance, hence such that it can be divided and destroyed; that the cake has sweetness and, say, a certain color, in and by itself, does not make it composite. For all I know, that might be a workable theological tactic; but I am not interested in theology here, nor am I interested in whether Anselm is guilty of this confusion. What I am interested in is that, after being tempted by the (controversial) tack of declaring different attributes tantamount to different parts, he *reasons* in a way that has no currency in analytic logic. He argues that (1) wisdom and goodness (and all other perfections) are the same thing and even that (2) either one of them (or any of the other perfections) is the whole of God, which, from an analytic point of view, does not make any sense. Wisdom is defined differently from goodness, so it is a different attribute; and any attribute cannot be the same as the substance it applies to—*however* that substance is understood: if, to get as close as possible to the landscape that seems to be inhabited here, a substance is understood as a bundle of attributes, then a bundle that includes the distinct attributes of wisdom and goodness cannot be the same thing as (one that includes) wisdom alone.

Does dialectical logic fare any better? In a way, yes, because in a monistic universe everything is the same, including God, his wisdom, and his goodness. But what is missing here is all the detail of how wisdom *turns into* goodness, or either

---
[2] Essentially the same problem and resolution are brought out in Anselm's *Monologion*, pp. 31–32.

of them, or both, turn into God. It is not as if there were any *other* kind of detail offered; hence inevitably any reconstruction of what Anselm was driving at will have to be based on conjecture; but the machinery of a narrative account and of narrative necessity is nowhere to be seen; and this absence, at the very least, should give us pause.

Let us stay with God a little while longer but bring him down from heaven, or perhaps have heaven itself come down and swallow everything else: make it/him into the totality of what there is—bottles and cabbages and humans. Let us look at how Benedict de Spinoza reasons. Only one thing (substance) exists, he claims, and that thing is God—though "God" is little more than a name for it and other names, like "nature" or indeed "substance," would be equally appropriate.

> *Except God, no substance can be or be conceived....* God is unique, that is ... in Nature there is only one substance, and that ... is absolutely infinite.... Except for God, there neither is, nor can be conceived, any substance ..., that is ... thing that is in itself and is conceived through itself. (1996, pp. 9–10)

So what about cabbages or humans? Here is where the qualification "oceanic" comes in. Suppose you are standing on the ocean shore, looking at the waves. You see one of them in particular looking powerful, running fast and tall toward you; you even identify for a moment with it, imagine it to be self-conscious and proud of its power and speed, to feel indestructible and immortal, only to be crushed a few seconds later and reduced to foam. You might react to the spectacle sadly and thoughtfully, and see in it a metaphor of the fate of human confidence and ambition; you might want to write that metaphor into an edifying story, using the natural spectacle you just witnessed to send a moral message down the way of your readers; but you might also become more matter-of-factly concerned with how (and how far) you can determine that what you were looking at was in fact *one and the same wave*. If you do, you will soon run into familiar problems: What are the exact points at which a wave begins and ends? How can I, with any level of precision, tell apart two waves (or is it just one?) advancing toward me? And, as you belabor such problems, it might dawn on you that perhaps you should be looking at the whole situation differently (give it a different meaning). There are, you might want to say now, no waves at all, if by that we understand individual, ontologically distinct, and independent things (if that is the meaning of "wave"); a wave is, rather, but a perturbation of the ocean, *a way in which the ocean is* (perturbed)—to use a Latinate word, a *mode* of it. That one cannot tell waves precisely apart, in this view, is a point of strength that confirms the view, not a weakness; the problems you were facing are not so much resolved as happily *dis*solved (it was an anomaly, not problems, you were considering). It is just right, and a welcome feature of your state as you now conceive of it, that you cannot quite say when a wave ends and another begins, for that is a telling sign that you should not overemphasize the multiplicity you are observing: that you should see the gradual merging of one wave into another, or into no wave at all, as suggestive of the fact that no actual, substantial contrast is to be implied there, that something is not a wave *to the exclusion of* another, *by denying* another—all that the talk of waves brings out is that *the ocean is (variously) wavy*.[3]

Take a particular cabbage now, or a particular human. We know that it is hard to maintain its identity in the face of all the difference the thing goes through during its lifetime, but focus now on the opposite issue: on the moment at which the analytic logician would finally breathe easily, because the thing is finally gone and difference can suddenly reassert its ontological role—when death, actual death, makes the autopsy recommended by analytic logic suddenly congruous. Gone, you might say, into what? And you might answer now, in a cheerful mood: into more of the same, into more of nature, or of substance or God, or of some other manner of referring to the undifferentiated blob from which everything comes, and into which everything eventually returns (Anaximander proposed *ápeiron*; I will discuss this later)—of which everything is just a way of being, a mode.

> By body I understand a mode that in a certain and determinate way expresses God's essence insofar as he is considered as an extended thing. (ibid., p. 31)

Once again, as with God's attributes in Anselm, difference does not create an impassable barrier, and it is not overcome by drawing ingenious connections. It is *absorbed* into a unity that denies difference as such, without bothering to resolve one difference at a time; that denounces the overstating of difference, because everything is ultimately a human or a cabbage, bald or not bald, a heap or not a heap, depending on how you look at it.

It is important to realize what these examples can do and what their limitations are. No one—not Anselm, Spinoza, or any of the other authors I will refer to later—has ever understood what he was doing as the playing out of an alternative logic. Spinoza thought of it as doing philosophy (ethics, to be precise, though I would call the above metaphysics) *more geometrico*, in the manner of Euclid—and Euclid was one of the greatest practitioners of analytic logic, the very archetype of an Aristotelian scientist, and a model of cogent (analytic) logical reasoning for 2000 years. Which made Spinoza look bad, for, despite all the paraphernalia of definitions, axioms, postulates, propositions, and the like that crowd his text and give it a superficial resemblance to Euclid's *Elements*, none of his proofs seems to work in the traditional, analytic, Euclidean sense—in that sense, they all seem to be failures and the resemblance is likely to be downgraded to a caricature. Hegel, on the other hand, appropriated Spinoza's substance (rechristening it "the Absolute"), as he tried to

---

[3] This last italicized formulation, in which "wave" has turned into an adjective, is more adequate than any I have used before to the new perspective I am illustrating (and equally, or even more, adequate would be adverbial formulations of the same content, as in "the ocean is *wavily*"). For in it (or in them), quite appropriately, the only noun contained in the sentence is "ocean," to be understood as a representative of the whole, which here will be the only thing there is. Nouns like "way" or like the Spinozian (and scholastic) "mode" are still deferential to traditional analytic logic, which thinks in terms of distinct substances and whose predominance has, not surprisingly, influenced the very structure of the language we speak, by having us employ a lot of distinct nouns that supposedly refer to those distinct substances—even when what we are speaking about is finding alternatives to that logic.

appropriate everything, and saw it as a rudimentary manifestation of the idea[4]; in his view too, then, Spinoza's reasoning is not to be accorded much independent respect. So what you need to do in order to see these as examples of what I intend to illustrate is be charitable to Spinoza and Anselm and all the others: think that such great minds must have been *after something*, and it couldn't be unsuccessfully attempting to prove theorems as Euclid did or functioning as a dumb sidekick for Hegel's more seasoned performance—something indeed that could be at the same level of depth, intelligence, and decorum as the logics of Euclid (i.e., of Aristotle) and Hegel; that could even provide an alternative, and a rival, to those other logics. By themselves, the examples will never *force* anyone to making this concession—that is their obvious, inevitable limitation—but perhaps adding more of them might stimulate general curiosity and benevolence.

My third example is Henri Bergson. A constant theme in his work is the contrasting of the neat divisions to be made in space with the radical unity of consciousness. Most relevant to us, this unity does not rule out a kind of plurality, except that all conscious episodes merge imperceptibly into one another—they are all one. We find this theme clearly expressed in his first major work (1910):

> We can thus conceive of succession without distinction, and think of it as a mutual penetration, an interconnection and organization of elements, each one of which represents the whole, and cannot be distinguished or isolated from it except by abstract thought. (p. 101)

> pure duration might well be nothing but a succession of qualitative changes, which melt into and permeate one another, without precise outlines, without any tendency to externalize themselves in relation to one another, without any affiliation with number: it would be pure heterogeneity. (p. 104)

And then we see it recur everywhere else in the corpus:

> The duration *wherein we see ourselves acting*, and in which it is useful that we should see ourselves, is a duration whose elements are dissociated and juxtaposed. The duration *wherein we act* is a duration wherein our states melt into each other. (1911a, p. 101)

> there can be no break in continuity between the child's delight in games and that of the grown-up person. (1911b, p. 68)

> Making a clean sweep of everything that is only an imaginative symbol, ... [the philosopher] will see the material world melt back into a simple flux, a continuity of flowing, a becoming. (1911c, p. 390)

Again this could be, as with Spinoza and Anselm, a metaphysical view, rather than a peculiar way of reasoning. They are all talking about special metaphysical

---

[4] "Corresponding to the concept of the absolute and to the relation of reflection to it, as expounded here, is the notion of substance in Spinozism. Spinozism is a defective philosophy because in it reflection and its manifold determining is an *external thinking*. The substance of this system is *one* substance, one indivisible totality; there is no determinateness that is not contained and dissolved in this absolute; and it is sufficiently important that in this necessary concept, everything which to natural picture thinking or to the understanding with its fixed distinctions, appears and is vaguely present as something self-subsistent, is completely reduced to a mere *positedness*" (1990, p. 536; translation modified). Note how the dissolution of determinateness that is characteristic of oceanic logic is regarded by Hegel as a *defect* in Spinoza.

4 Oceanic Logic

entities—God or consciousness—which we might expect to have special metaphysical characters. But note that, in his last major work, Bergson applies the same attitude to something less unique and exceptional. In (1935) he has identified two brands of morality and religion, and the question arises of what authorizes him to call these quite different things by the same names. In the case of religion, he answers by making a virtue of a slippery slope:

> we find interposed ... transitions and differences, ostensibly of degree, between two things which are as a matter of fact radically different in nature and which, at first sight, we can hardly believe deserve the same name. The contrast is striking in many cases, as for instance when nations at war each declare that they have God on their side, the deity in question thus becoming the national god of paganism, whereas the God they imagine they are evoking is a God common to all mankind, the mere vision of Whom, could all men but attain it, would mean the immediate abolition of war.... [T]here is a difference between the two cases, and if we take it into account, we shall notice, in the matter of religion, a gradual disappearance of the opposition. (p. 183)

Bergson's masterpiece was *Creative Evolution* (1911c); there he argued fancifully for a life that asserts itself against all obstacles—indeed, that posits obstacles for itself which it then must overcome, and does.

> The impetus of life ... consists in a need of creation. It cannot create absolutely, because it is confronted with matter, that is to say with the movement that is the inverse of its own. But it seizes upon this matter, which is necessity itself, and strives to introduce into it the largest possible amount of indetermination and liberty. (p. 265)

Note however that evolution, and change in general, can be read by stressing two distinct aspects of it. One is identity in difference: the fact that something stays the same while also acquiring new properties. This is the aspect stressed by dialectical logic. The other is the fact that change happens with infinite graduality: however minutely you divide the phases it consists of, you will always find in each phase some sub-phase in which the thing is still the old way and some later one in which the new way has already turned up.[5] The two ways, in Bergson's language, "melt" into one another, and this language shows that he is not looking at the phenomenon in a dialectical way—he is focusing not on conceptual articulation but on what can*not* be articulated, not on how detail and distinctiveness are gained but on how they are lost.

> The mistake of ordinary dualism is that it starts from the spatial point of view; it puts on the one hand matter with its modifications in space, on the other unextended sensations in consciousness. Hence the impossibility of understanding how the spirit acts upon the body or the body upon spirit.... We have striven to show that this psychology and this metaphysic

---

[5] When it comes to empirical (for example, biological) evolution, the conjunction of these two aspects generates some tension. Say that you are looking at how a fin evolves into a wing. Both a fin and wing are selectively advantageous; hence we can understand how they were both retained; but how did we get from one to the other? How did selection retain all the infinite degrees of the evolution from the one to the other, most of which were probably *dis*advantageous? So we are not surprised to see phyletic gradualism (according to which evolution is smooth and continuous) give way to Stephen Jay Gould's punctuated equilibrium, which has evolution proceed in fits and starts. Once again, part of what is happening here is that one logic gives way to another.

are bound up with each other, and that the difficulties are less formidable in a dualism which, starting from *pure* perception, where subject and object coincide, follows the development of the two terms in their respective durations—matter, the further we push its analysis, tending more and more to be only a succession of infinitely rapid moments which may be deduced each from the other and thereby are *equivalent to each other*; spirit being in perception already memory, and declaring itself more and more as a prolonging of the past into the present, a *progress*, a *true evolution*. (1911a, p. 120; last emphasis mine)

My last historical example (for now) comes from Martin Heidegger's (1962). Early on, when introducing Being-in-the-world, Heidegger provides the following characterization of it:

> The compound expression "Being-in-the-world" indicates in the very way we have coined it [by hyphenating it, which is also true of the German "*In-der-Welt-sein*"], that it stands for a *unitary* phenomenon. This primary datum must be seen as a whole. But while Being-in-the-world cannot be broken up into contents which may be pieced together, this does not prevent it from having several constitutive items in its structure. Indeed the phenomenal datum which our expression indicates is one which may, in fact, be looked at in three ways. If we study it, keeping the whole phenomenon firmly in mind beforehand, the following items may be brought out for emphasis: First, the "*in-the-world*." With regard to this there arises the task of inquiring into the ontological structure of the "world" and defining the idea of *worldhood* as such.... Second, that *entity* which in every case has Being-in-the-world as the way in which it is.... Third, *Being-in* as such. We must set forth the ontological Constitution of inhood itself. (pp. 78–79)

Heidegger insists that the "constitutive items" that make up Being-in-the-world are not attributes attaching to a substance. That would make them *categories*, which they are not; they are *existentialia*.

> What is meant by "*Being-in*"? Our proximal reaction is to round out this expression to "Being-in 'in the world'," and we are inclined to understand this Being-in as "Being in something" [notice the lack of hyphens]. This latter term designates the kind of Being which an entity has when it is "in" another one, as the water is "in" the glass, or the garment is "in" the cupboard. By this "in" we mean the relationship of Being which two entities extended "in" space have to each other with regard to their location in that space.... "Being-in," on the other hand, is a state of Dasein's Being; it is an existentiale. So one cannot think of it as the Being-present-at-hand of some corporeal Thing (such as a human body) "in" an entity which is present-at-hand. (ibid., p. 79)[6]

Whatever sense this makes (and assuming it makes some—which is not for me to decide here), it cannot be captured by analytic logic. The "state" that Being-in is of Dasein's Being is not detachable from it. Every *existentiale* which articulates the ontological structure of Dasein is a way of looking at that whole structure (keeping it all firmly in mind) and bringing out some of it "for emphasis." It is a point of view on it, a mode, or way, in which it presents itself to us when we look at it from that

---

[6] Heidegger has a strong inclination to making up new words for what he takes to be new content—an inclination that we have seen to be typical of analytic logicians (see footnote 11 in Chap. 2 and the attending text). His intuitions may be veering away from analytic logic, but his linguistic habits are still informed by it. Which is to be taken not as a criticism but rather as evidence of something I will focus on in Chap. 11: All three logics are always in play together in our intellectual endeavors. Similar remarks could apply to Anselm, Spinoza, and Bergson (and importantly qualify what I said above on Spinoza's commitment to Euclid).

point of view (much as my profile is a mode of my being: a way you see me when you look at me from a certain angle), not a reality that might fall out of it and go on to lead its own independent life (as my profile would be if you could detach it from me, or a smile if you could detach it from the Cheshire cat)—perhaps to become a "constitutive item" of something else. Analogously, an individual person can respond differently to different companions and even be taken as a consequence to be different people (to be schizophrenic, divided unto herself); but then, when you consider the matter closely, you figure that those different responses come from the same place—that all that is different about them is how they are differently resonant to their various perspectives, how they provide variations on the same theme.

Would dialectical logic, again, do any better? Note that here, too (as in previous examples), there is nothing of what Hegel called "the labor of the negative"[7]: the toilsome superseding of contrariety through the tracing of ingenious connections. There is just a glance that moves from one side to another of an integral reality, never feeling any stress as it illuminates different angles of it—different angles of *one and the same thing*, which is there for everyone to appreciate. What stress is felt comes from opposing any misunderstanding of that reality: crucially, misunderstanding its integral character and reducing it to bits that would then have to be left irreparably alone (analytic logic) or painstakingly stitched together (dialectical logic). What emphasis is voiced is to the effect that the reality was *always already* integral: that its integrity was always already compatible with its having a complex, interesting architecture, and that the complexity of the architecture follows from the various ways you can choose to relate to it.

I mentioned the fact that my examples of (what I take to be suggestions of) oceanic logic in the history of philosophy deal with exalted matters: God and the universe and consciousness. But, if oceanic logic is to have to a chance to be on a par with its rivals, it must hold its own in more mundane circumstances; so, as a premise to moving from examples to doctrine, let me consider one such mundane case, where a group of people (a city council, say) are deciding on some policy to be enforced and the values being thrown around by them are safety on one hand and freedom on the other. An analytic attitude to the debate would be to unambiguously define "safety" and "freedom," clearly state their opposition, and then, by bringing out their respective pros and cons and comparing and contrasting them, become a staunch defender of the one or the other (but, crucially, not both). A dialectical attitude would judge one of the two values (e.g., safety) to be more mature than the other, a later stage of dialectical development, and argue that people certainly want, and have a right, to be free, but the enjoying of that right demands that they be able to walk the streets, patronize stores, and have their say on any public matter, safely, so let us work on safety, as the necessary actualization of the promise involved in freedom. An oceanic attitude would amount to claiming that, when people talk about safety or freedom, they are talking about the same thing: they are foreground-

---

[7] "Thus the life of God and divine cognition may well be spoken of as a disporting of Love with itself; but this idea sinks into mere edification, and even insipidity, if it lacks the seriousness, the suffering, the patience, and the labor of the negative" (1977b, p. 10).

ing two angles of the same desirable state—describable as a calm, assured fruition of one's due or as the self-expression that fruition elicits.

In the course of making her point, an oceanic logician will make moves that remind us of dialectical ones. To rehearse an earlier illustration, she might say things like: How can someone be free if she cannot go about her business in a secure way? But what matters is that there will be no asymmetry to her reasoning, for sooner or later she will turn around and make the converse point: she will say now that someone is not really secure if that "security" is bought at the price of intimidation, or unless what she is "secure" doing is a (free) manifestation of her potential. Which means that, in practice, she will be driven to solutions for which common parlance has a name typically taken as derogatory: she will reach for a *compromise*. That is, in a more sympathetic interpretation of that word than is common, she will look (sometimes quite creatively) for some policy that best substantiates the *single* value of which both freedom and safety (she has argued) are "constitutive items." She will aim at an end result in which "everybody (or, better, everything) wins," because "everything" is what she has always been talking about, under the guise of talking of freedom or security.

Compromise is a practice with as much practical prominence as it receives intellectual disdain. Politicians, we know, diplomats, and merchants are good at their job to the extent that they negotiate effectively and are able to get out of tense confrontations by engineering compromise solutions that will be acceptable to all parties. Which, in worst-case scenarios, is regarded as a swindle ("the art of the deal") and, in slightly less unfortunate ones, as everyone giving up something of what they want in fear of doing worse. The reason for such disdain must be made explicit, on the way to rejecting it: The forms of reasoning explicitly recognized so far see, and offer, no genuine way of resolving a confrontation, tense or otherwise. What is recognized as possible by these forms of reasoning is: one party wins outright (the preferred analytic option), or one party colonizes (a more subtle kind of winning over) the others, showing them how to graduate into itself (the preferred dialectical option), or, for lack of anything better, all parties agree to some measure of loss[8] and then, since the confrontation has not in fact been terminated by this move, each party waits until the next occasion to stick it to the others. None of this does justice to political action in general, and it all fails miserably to account for the real manifestations of genius in it—Martin Luther King giving his dream speech, say, or Sadat going to Jerusalem, or Mandela instituting the Truth and Reconciliation Commission—thus proving that the failure in question here is an intellectual one:

---

[8] A philosopher who has paid considerable attention to the concept of compromise (which he labels "bargaining") is David Gauthier, so we can use him as an example of what compromise looks like from an analytic point of view—in confirmation of the claims made in the text. In Gauthier (1986) we find the following description: "The procedure of bargaining may be divided into two principal stages. First, each party advances a claim—proposes an outcome or joint strategy for mutual acceptance. In general the claims of the parties are incompatible. Hence second, each party—or at least some party—offers a concession by withdrawing some portion of his original claim and proposing an alternative outcome. Barring deadlock the process of concession continues until a set of mutually compatible claims is reached" (p. 133).

that we have not yet recognized, and worked out, the type of *reasoning* that goes on in such cases, the *logic* that drives them; so political action had to proceed as best it could base on intuitions that were often sound but were also sneered at by impotent theories of the *logos*, whose sneer reflected their impotence.

Let us get to it then, and, to begin with, let us combine the political flavor we began to taste in the last few pages with classical sorites strategy. Let us consider two slight variants of the Bald Man and the Heap. The first one is the White Man: a man whose skin is absolutely white (or maybe pink) is white, and if you make the skin of a white man darker by an infinitesimal degree, the result is still a white man; therefore all men are white. The second one is the Heap of Trouble, where a heap of trouble is defined as an amount of trouble a community cannot handle: a perfectly calm situation is not a heap of trouble, and if you add an infinitesimal amount of trouble to a situation that is not a heap of trouble you do not make it a heap of trouble; therefore there are no heaps of trouble. Suddenly what were regarded as paradoxes can be seen as making legitimate, even important, points. That there is no sharp line dividing whites from non-whites is no longer an embarrassment: it becomes a *reason* (among several, of course) for not discriminating between the two, as any such discriminating behavior would have to be based (among other things) on a purely arbitrary *logical* choice—an arbitrary choice of how exactly to determine the logic of "white": its contribution to meaningful discourse. That there is no sharp line dividing what is too much trouble from what is not becomes a reason (again, among several) why a community should never look for the easy alibi that some situations simply cannot be handled and hence should always demand and expect more of itself. Which is not to say that there will not be contexts where it is appropriate and opportune to distinguish between a Category 4 and a Category 5 hurricane, and to do so on the basis of just the sort of choice (of specific wind speeds) I have called arbitrary, as such generic classifications can be of help in soliciting appropriate and opportune reactions on the part of citizens and institutions (hence analytic logic is also going to be relevant to this matter, as I will argue in Chap. 11); but it *is* to say that no citizen or institution will ever be justified in saying to themselves, "There is nothing I can do *because* I am facing a Category 5 hurricane." Here Eubulides will provide effective assistance, alerting us to the fact that there is only a difference of degree, not of substance, in moving from a maximal Category 4 to a minimal Category 5 storm.

In general, any situation that would traditionally be regarded as an instance of the *paradox* of sorites, and hence as causing embarrassment, is a natural case for oceanic logic.[9] The tendency here, as in dialectical logic, though for different reasons, is to go all the way toward monism and think (with Spinoza, and possibly Bergson) that everything is the same; but, as in dialectical logic, that tendency can be inhib-

---

[9] On the other hand, the rigid distinctions associated with mathematical entities and with other entities modeled on them constitute an anomaly for oceanic logic. (As was suggested in footnote 14 in Chap. 2, this anomaly will be taken up in Chap. 9.) Ambiguities will be perceived here in a way that is similar to dialectical logic—as challenges to come up with appropriate sorites—and so will spacetime continuants (more about the latter in Chap. 6, in the discussion of Zeno's paradoxes).

ited and local applications of oceanic logic are perfectly legitimate. It will be legitimate, for example, to just consider humans and then claim that every human is white or black, depending on how you look at her—*as a fitting sorites proves*. What is typically taken to be a black person, because she is compared with whiter shades of pale, can be judged to be white if we run the spectrum in the opposite direction; and the same is true for what is typically taken to be a white person. (If the reader did her homework earlier, and proved *dialectically* that Bruce Springsteen is black, she will be able to prove it again—but in a different style of reasoning.) So oceanic logic will make confrontations and contraries melt away, not by reducing everything and everyone to a bland lowest common denominator or by telling self-satisfied just-so stories[10] but by exploiting the porousness of reality, the flexibility of borders, and the sudden familiarity the unknown acquires when you move toward it. Everything will recognize itself in its other not because it will see the other as a phase of its own development but because it will see the road between itself and the other as one that can be traveled without impediment and in both senses. The world, and everything in it, will continue to be an exquisite and opulent place, where there is white, black, pink, and brown, and indeed *everything* is white, black, pink, and brown: it has it in itself, as part of its own logic, to be all those things. And compromise will be the natural way of addressing an issue, not as a practice of grudgingly conceding what cannot be obtained but as one of intimately resonating to all the difference there is.

Oceanic logic cannot be formalized either, which accounts for the relentless annoyance experienced in trying to come up with a(n analytic) formal logic of vague predicates. The way *pink* melts into *red* has nothing in common with the way *innocence* melts into *guilt*, or an episode of my conscious life melts into another, or everything there is melts into God: each of these sorites "works" in a way that reflects its specificity and cannot be adapted to any other. As with dialectical moves, we can uninformatively summarize what we did in a particular sorites by saying that we proved some predicates to gradually merge, which, as within dialectic, will not help when we try our next case. Here, too, we must remain focused on all details of the world, as any such detail can be of logical significance. If there is a distinctive, singular character to analytic logic, it is that only in it we are allowed to lose track of most of what there is. And it is to be regretted that, in taking analytic logic to be the only theory of the *logos*, we have also taken logic to unquestionably imply worldlessness.

There is a flip side to oceanic logic's worldly character, which can be gotten at by finally asking what *meaning* amounts to in it. Whether this logic is applied globally or locally, the meaning of a word is the totality of the context of application: both "white" and "black" mean every human, if it humans that we are talking about, as "wave" means the whole ocean—white and black and waves are just aspects of the whole thing, whose destiny is to merge into any other aspect. And mind you: not

---

[10] As pointed out before, dialectical logic works retrospectively, by telling a plausible story about what has already happened, by *re*constructing it rationally. So the reference to just-so stories is not unfair, as in both cases we give ourselves the benefit of 20–20 hindsight.

the whole thing in the sense that is pertinent to dialectic, of a narrative that captures every such aspect and finds a truthful role for it. No: the whole thing insofar as each aspect is a limited point of view on it and redeems its limitation only by losing itself in others (that is what merging is), hence proving the delusiveness of thinking that there was something substantive to it—that it was anything other than a mode of substance. Which returns us to God and shows that, for all our bringing this logic down from heaven into mundane dealings, an important element of the unintelligible remains vital to it.

Analytic and dialectical logics do their work by emphasizing and exploiting the immense power of language, either in carving nature at its joints or in sealing all cracks in nature with the wax of plausible narration. Both display an ambition of reducing the world to something akin to language. Oceanic logic displays the opposite tendency: not just God, the whole world, or life but any small matter to which this logic is applied exposes the limitations of language, its making distinctions without a difference, and the need not to get sidetracked into those distinctions—into language at all—if we are to relate usefully to matters large and small. The good motivation that invites us to reach for a compromise—as contrasted with all the bad ones arising from cowardice, disingenuousness, and manipulation—is awareness of how little each of us sees and how much better it is to project a common vision. But, however common that vision, it is still limited, and the world still looms distant and unfathomed; oceanic logic predicates its own, and language's, failure to reach it. Both the precision of a mathematical definition and the seductiveness of a well-told story are anomalies for it, insofar as they deny that failure, which may be a main reason why this logic has not been recognized so far as a worthy alternative to its two rivals: doing logic, as much as doing anything else, typically goes together with having a proud conception of what you do—specifically, here, of how distinguished and decisive the structure of the *logos* is, how revealing of its object. And yet, even what exposes the ways in which meaningful discourse systematically fails to attain its goal is a doctrine of the *logos*, and even arguments enlightening us on those failures are forms of reasoning.

Interlude: A Conversation Between Two Oceanic Neighbors

Jack: Hi Don. How are you doing?
Don: Just the person I was looking for. I need to talk to you.
Jack: What about?
Don: It's about our backyards. See, it's been a while since we moved in…
Jack: A couple of months.
Don: Indeed, and so far we have been leaving everything open. Anyone can just wander back and forth between your property and mine.
Jack: Is there anything wrong with that?
Don: Yes, I think a lot is wrong with that. We should build a fence and share the expense.

Jack: That's some news. Why would I want to do that? I like the open space. I like the view. Why would I want to shut myself up in a cage?

Don: What? Don't you see you are in a cage right now?

Jack: No, that's exactly the point: I am not. When I come out my back door I can see as far as I want, with no wall blocking my view. I can see my plants and yours, I can count the lemons on my tree and the oranges on yours. I can even see you, and mostly I enjoy that—except when you start talking about building a fence.

Don: Jack, have you been at the zoo, lately?

Jack: Not for a while. But what does the zoo have to do with anything?

Don: Even if it's been a while, you will remember the aviary.

Jack: How could I forget that? It's the most beautiful exhibit there: all sorts of greenery and rocks and waterfalls, not to mention all the birds flying here and there, or perching on a branch, birds of incredible colors, some with the strangest heads and feathers and beaks.

Don: I would imagine that the birds feel as if they were free to fly, or to perch, anywhere?

Jack: Yes, I would think so.

Don: But in fact they are in a cage.

Jack: I see what you mean. The aviary is enclosed by a plastic dome, but that is so high up and so large that the birds will never, or at least very rarely, become aware of it.

Don: I am sure you are right, and I am sure that, if and when they do, it won't bother them too much. They will think of it as just another obstacle in their path, and simply change directions.

Jack: Once I had this bird that for months was trying to get in through one of my windows, and went on forever pecking at it. The poor thing could not understand how something as transparent as air—something that for it *was* air—would not give in to its flight as air does. Eventually, it changed directions.

Don: OK, Jack. But that's birds. I just said that they will "think" of it, but that was not the right word to use. Birds don't think; so they are in a cage but do not realize it. Humans, however...

Jack: ... are thoughtful, reflective beings. So what?

Don: So, when they are in a cage, they will come to realize it.

Jack: Yes: if I were to lock you in an aviary, sooner or later you would become aware of the plastic dome and realize that you are caged in. I still don't see where you are going with this.

Don: You will soon enough. So you judge yourself free because you can see my plants and my oranges, and can see me when I come out in my shorts and lie in the sun?

Jack: It's a beautiful sight, let me tell you.

Don: I am glad you like it. But, from where you are, you must also see the fence I have on the other side of my property, blocking your view from penetrating my other neighbor's property.

Jack: Yes, you guys sorted out that issue very quickly.

Don: He was not as eager as you are to look at me in my shorts. But the point is: though you don't have a wall here, you have one there. And you have a wall on the other side of *your* property, blocking your view of the street.
Jack: The city put that in.
Don: Nothing to argue there. And you have walls and fences everywhere else you look in the street. What you can actually see is very limited.
Jack: I can still see a lot. I can still walk everywhere.
Don: It depends on how you think of it. You can think of yourself as being able to go everywhere and see everything, *except for* some areas that are blocked, as they belong to someone else who has decided to wall himself in. *Or* you can think of yourself as locked into a particular path: forced to walk on it only and to see only what is accessible from it.
Jack: Yes, I agree that I can think of myself in either of those two ways, though I would add that the second way of thinking is far more depressing than the first one, and I would not want to indulge in it. But what about us now?
Don: Well, what difference does it make to you whether we build a fence between our properties, when you are caged anyway?
Jack: It makes a lot of difference, because without this fence of yours I can see a lot *more*, and feel a lot freer.
Don: And how much should you trust that feeling? Do you believe that someone sentenced to life in prison would be better off if he was given a larger cell, even one that was ten times larger?
Jack: He would be more comfortable.
Don: I don't deny it. But what I am asking is: would he be better off *as far as his freedom is concerned*?
Jack: Probably not.
Don: That's right. Do you think it makes any difference for Truman Burbank, of *The Truman Show*, when he finally discovers his plastic dome, how large his cage is or how comfortable he has been living in it?
Jack: OK, I get it. But, given that I am caged no matter what, I might as well have a more spacious and comfortable cell.
Don: We will get to that. For the moment, let me just say that, as you are always caged, you are also always free.
Jack: You are all twists and turns. You've got to explain this now.
Don: It's not so hard, because again it's all a matter of how you think of it. It's just as I said before: you may think of all the movement you are allowed in this world as enclosed within fences and walls, as constrained by them no less than the walls of a jail constrain what a prisoner can (and, mostly, *cannot*) do; or you may think of all the fences and walls as rules that define your free movement, conditions of it, and think of your movement as remaining free and totally under your control within the space you are allowed, while abiding by those rules and conditions. For you are free to walk one step, and another one, and another one yet; and when you come before a wall you may think of yourself as blocked by the wall or, instead, as free to move laterally, away from it. None of this is going to change, no matter how far

Jack: the wall is. It's always going to be dependent on how you think of it, on how you think of yourself.
Jack: So the prisoner can also think of himself as free?
Don: Of course, and yet his condition, whether he sees it as free or caged, can also be more or less convenient, more or less comfortable. But at least we've got the whole issue of freedom out of the way.
Jack: Let us talk about convenience and comfort, then, for the fence doesn't promise any of either to me.
Don: You said that one way of thinking of this is far more depressing than the other; so, since whether you are free or not depends on how you see yourself, you might as well "indulge" in seeing yourself as free and determine what is the best way of organizing your free space.
Jack: The best way to arrange my cell.
Don: If you prefer. In your own house, you find it more convenient and comfortable to have separate rooms, where you can attend separately to the various tasks of your everyday life, in ways that are more efficient and also more—how shall I put it?—"appropriate," rather than having, say, a single gigantic space, like a big tent, where all different activities are performed together: cooking and sleeping and listening to music and taking a shower and defecating.
Jack: That would be peculiar.
Don: It would, wouldn't it? Same here. Forgetting about freedom, which is a red herring, let us ask ourselves a factual question: is it more convenient and comfortable to leave all this space open, or wouldn't it be better to divide it up according to our different interests, so that you can visit when you want to but also do not have to stand my presence, or I yours, when we have other business to take care of?
Jack: And the answer to this rhetorical question is?
Don: But, clearly, it's better to divide up the space, so your dog can be better controlled...
Jack: Why does Immanuel need control?
Don: Not for my sake. But sometimes I have guests, and they might not like it; and sometimes you might have guests, and you might want to have your privacy.
Jack: Wait a minute, Don; I think you're playing a trick on me.
Don: How so?
Jack: You have put the issue of freedom aside as being a matter of point of view, of perspective, and you have described the issue of convenience and comfort as a factual one that can be decided objectively.
Don: Is there a problem with that?
Jack: Yes, there is. Because what you said of freedom applies to comfort too. Indeed, it seems to apply to pretty much anything at all.
Don: You don't say!
Jack: But I do. Imagine I am sitting on the couch, reading a book. Everything is fine, there is harmony in the universe, except for this tiny bump in the cush-

ion underneath that is bothering me ever so slightly. I could get up and straighten up the cushion, which would interrupt my leisurely and pleasant reading, or I could just go on reading, judging the nuisance too small to worry about. The point is: I have to decide how to think of my situation. If I think of it as uncomfortable, I will get up and fix it. But I can also think of it as *not* uncomfortable, and then I will do nothing. Here, too, it's a matter of perspective, of point of view.

Don: That, however, is a decision about a very marginal case of discomfort.

Jack: Yes, but how wide is the margin? Imagine that the bump gets larger and harder, and maybe sharp too. Imagine that a fly starts buzzing in my ear. When will I decree that this is enough discomfort, or even that it *is* discomfort at all, as opposed to just a lower degree of comfort, so that I get up and do something about it?

Don: I agree, the margin is a bit loose. You think that the same is true of our case?

Jack: Well, suppose I was living in a big tent, and I moved to one corner or another of it when I had to cook, or sleep, or play music, or defecate. If I started thinking about it, I might find the situation less than ideal; but how is that different, in principle, from being bothered by the bump in the cushion, or by the fly? Isn't it up to me, here too, to decide whether I will judge the situation uncomfortable? And to decide whether I will do something about it? Isn't this, also, a matter of perspective, of point of view?

Don: I see. So, when it comes to our backyards...

Jack: It's the same thing. I admit that were my dog to wander onto your property when you have guests who hate dogs, or were you (or I) to walk around naked, the situation might be less than ideal. But what forces us to call that *uncomfortable*? "Comfortable" and "uncomfortable" are simple opposites, and simple opposites make life sound too easy. What there is in life is an infinite gradation of nuances, and whatever nuance might apply to our situation it will never, in and by itself, determine which of those opposites we have to use. It will always be how we see the situation, how we choose to see it; and we could always choose to see it differently. Which means: we can always choose to *act* differently, depending on how we see it.

Don: The more I listen to you, to us, the more I become convinced that we are both talking about the same thing and expressing the same general interest, but we have been focusing on one or another aspect of it, one or another angle, so we end up giving the impression of disagreeing, giving that impression even to ourselves, when we could easily shift into the other guy's perspective and see our positions as complementary, not as conflicting.

Jack: How do you mean?

Don: You have been stressing freedom and I have been stressing comfort. But these are two names of the same thing. You want to feel comfortable on the couch, reading your book, because you want to be free to devote yourself to that favorite pastime of yours, and whatever limits your comfort also limits your freedom. And you favor that pastime because it makes you feel at home with yourself and the world, and whatever limited your freedom (if, for example, you had to rush to a meeting) would also limit your comfort. I want to be comfortable in my space because I want to be free to move in it as I like, and you want to be free to look in the distance because anything that blocked your view would make you feel uncomfortable. So the right thing to do would be to try and each obtain the maximum amount of freedom/comfort we can manage together.

Jack: Now you're talking. How could we obtain that?

Don: Not easily. If all I had to do was to prove you wrong, I could just poke holes in your arguments, except that you would also poke holes in mine and we would never be done with it. But once we realize that, although we started from different points, we are aiming at the same thing, it gets hard: we must think hard of how to get there.

Jack: I have an idea. Why don't we build a retractable fence, one that we could pull all the way to block the whole view or push back all the way to leave the view open, or even pull part of the way if that's what we want, and then we would be able to accommodate the fence to our different desires and requirements, day after day?

Don: You mean built like an accordion, with slices that fold?

Jack: Yes, that might be the best way, or we could have it slide on a rail and disappear between our two houses.

Don: That might be expensive.

Jack: So we have to do some research, and then we will have to decide if we both judge the outcome to be comfortable enough.

Don: Yes. To be honest, I find it very uncomfortable to spend too much money on it. It would make me worried, unable to enjoy my leisure time.

Jack: That's right, and if that were the case then the outcome would also not be liberating, because you (and I) would be constrained by our worries, forced to obsess about them.

Don: We have to find the right compromise among all these factors, where every factor is given just the right weight.

Jack: So that every factor wins its battle, which can only be won by having every other factor win.

Chapter 5
Necessity

Abstract A logic is a normative theory, which states not how people *do* reason but how they *should* reason; so, in further articulating the differences among the three logics, it is best to start by considering how differently *necessity* plays out in them.

In analytic logic, the necessity that sanctions an inference is derived from the one by which contraries exclude one another: if "mortal" and "immortal" are contraries, then once Socrates is subsumed under mortal humans, its link with mortality can never be severed. An analytic proof may require tremendous ingenuity; but if the right premises are found and the right inferential steps are made, this outcome appears set in stone (and independent of the heuristics of it).

The necessity ruling dialectical logic is the narrative kind: it amounts to redescribing a concept (say, being) so that it is shown to coincide with a contrary one (say, nothing), much like a good narrator redescribes a character, an event, or a situation so as to make it look natural that a momentous change would occur in them *while they remain the same thing*. This necessity seems looser than the analytic one but that impression is based on a misunderstanding: analytic logicians divide up the problem of providing logical accounts of ordinary arguments by first offering arguments in artificial languages which, being artificial, can be defined with total accuracy and then translating from the artificial languages into the ordinary one—at which point all the looseness that was first put aside resurfaces. Dialectical logic, not being formalizable, cannot divide up the problem in this way, so the looseness is obvious throughout.

Whereas the necessity current in both analytic and dialectical logics is brought out by attention to detail, the one current in oceanic logic blurs detail and indeed makes us think of detail as confusing: as making us miss the forest for the trees. Similarly, the identity current in this logic undervalues intellectual, conceptual qualifications and stresses the undifferentiated, material stuff that is to be found at the basis of every alleged distinction.

As previously announced, in the next several chapters I intend to highlight individual features of the three logics, which in most cases have already been mentioned but need to be carefully examined in order to bring out the specificity of these reasoning and argumentative tools. Since a logic is a normative theory, it might be a good idea to start with what makes it normative: its being intrinsically implicated with modality; the *necessity* that qualifies its judgments.

Consider the most famous example of Hegelian dialectic: the master-slave section of (1977b).[1] The slave is dependent on the master's will, which is, on the other hand, independent of the slave's very existence: if that particular slave were to disappear, or never to have been, the same will would be exercised on other unfortunate humans, with the same consequences and enforcing the same structure. Whereas the slave would quickly lose his life if the master decided to deprive him of food and shelter or if he, quite simply, willed the slave's death. So there is a clear direction in the hierarchy of authority and submission here, with the master on top and the slave on the bottom. And yet, what would a master be without a slave? An insane individual addressing empty commands to no one and issuing the dictates of his inept will without arousing anything more than a compassionate, or sarcastic, smile. What, or who, if not the slave, makes the master *a master*, defines him as such (in more Hegelian terms: defines the nature of his consciousness), and hence can invalidate the definition by simply refusing to collaborate—by accepting death, if need be, as long as the master remains surrounded by an unresponsive environment?[2] The person who happens to be a master might well be independent of the person who happens to be a slave; but *the master in him* is essentially dependent on the slave assuming his role as a slave—no slave, no master—and the decision of assuming that role belongs to the slave only, independently of the master. Therefore each of the two, the master and the slave, is at the same time dependent on, and independent of, the other.

One might appreciate the ingeniousness and resourcefulness of this reasoning and still find it inappropriate that it be described as providing a necessary link. Say that in Euclidean geometry we prove the sum of the interior angles of a triangle to be 180°; *that* conclusion, one might tell us, really shows up as necessary, as imposing

---

[1] This is Section A of *Self-Consciousness*, pp. 111–119. The following passages are especially relevant for the discussion that follows: "The lord is the consciousness that exists *for itself*, but no longer merely the concept of such a consciousness. Rather, it is a consciousness existing *for itself* which is mediated with itself through another consciousness, i.e. through a consciousness whose nature it is to be bound up with an existence that is independent, or thinghood in general.... The lord relates himself mediately to the bondsman through a being [a thing] that is independent, for it is just this which holds the bondsman in bondage; it is his chain from which he could not break free in the struggle, thus proving himself to be dependent, to possess his independence in thinghood.... [T]he lord achieves his recognition through another consciousness.... Here, therefore, is present this moment of recognition, viz. that the other consciousness sets aside its own being-for-self, and in so doing itself does what the first does to it. Similarly, ... this action of the second is the first's own action; for what the bondsman does is really the action of the lord.... [W]hat the lord does to the other he also does to himself, and what the bondsman does to himself he should also do to the other.... But just as lordship showed that its essential nature is the reverse of what it wants to be, so too servitude in its consummation will really turn into the opposite of what it immediately is; as a consciousness forced back into itself, it will withdraw into itself and be transformed into a truly independent consciousness" (translation modified).

[2] This is the sense in which a strike, whatever the negative economic, legal, or even physical consequences of it for the participants, shows the workers acquiring consciousness of their power.

5 Necessity 63

itself on our assent without leaving us any alternative option and any room for disagreement. But this other reasoning—if we are even entitled to call it "reasoning"—draws whatever force it has from projecting onto the matter a highly idiosyncratic point of view and from redescribing the matter in a way that is perhaps convincing once you phrase it but that one does not feel forced to accept or to phrase that way. Who knows how many other redescriptions are available, and what other surprising turns we might take if we adopted them; but *why* should we adopt them? At best, we could say that the joint dependence/independence of both master and slave is a *possibility*, not that it has the firm necessity of the Euclidean theorem.

There is no denying that necessity acquires different meanings in different logics, and part of what I need to do here is unpack this difference. But, for that task to be fulfilled and for the contrasting senses of necessity to emerge clearly, we must first dispel other apparent differences that crowd our understanding of the situation (and are largely responsible for the vibrant, and confused, protestations above), as a result of the long favor analytic logic has enjoyed and of the extent to which, as a result, it has appropriated value judgments it does not quite merit. I will take up this dispelling now, in order to clear the ground for the determination of what in fact the contrasting sorts of necessity are in the various logics.

One alleged difference is perhaps not so much apparent as misleadingly described, for we know that the construction allowing us to prove the Euclidean theorem about the interior angles of a triangle amounting to 180° is no less gratuitous, given the theorem's premises and conclusion, than is the "projection" allowing us to dialectically develop the concepts of master and slave. We are, for example, to extend one side of the triangle and then draw through the same vertex of it a parallel line to the opposing side, and recognize the two exterior angles thus formed as either alternate or corresponding, and in either case equal, to the two interior angles on the opposite side, and finally point out that the sum of these three adjacent angles (the interior one and the two exterior ones) is equal to two right angles.

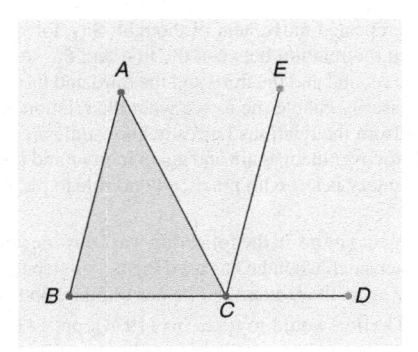

Where does all this extending and drawing and recognizing come from? Doesn't it also amount to redescribing the situation in ways that were not predictable on the basis of a simple inspection of it (as is portrayed in the statement of the theorem) and certainly cannot be considered *necessitated* by that very situation? In retrospect, it works beautifully and persuasively; but doesn't the Hegelian redescription, retrospectively, work just as well?

What is different here, as mentioned in Chap. 3, is that the redescription, and hence the construction, substantiating the proof is not supposed to belong to the logic of the situation but to the heuristics by which we uncover that logic. Once the logic is uncovered, we can let the heuristics go and record the outcome in terms of pure logical relations. If we were to take the extreme position recommended by Descartes in the 11th of his *Rules for the Direction of the Mind*,[3] we could regard that construction, and indeed the whole proof, as nothing other than crutches (or, if you will, as a Wittgensteinian ladder[4]) sustaining our faulty, limited mind, which is supposed to run through the construction and the proof at increasing speed until it gets to the point of *seeing* the logical relations outright and triumphantly throwing off the crutches (or ladder). But, even without going that far, constructions and diagrams will retain an uneasy role, and an uncertain reputation, within the pure confines of geometry: auxiliary implements, visual tools, illustrations—not the real thing.[5] And we know why that is so, by now: because "the real thing," here, is to be established once and for all and then exported, identical with itself, to any other context where it might be found relevant; so, once we have it, we might as well forget how we acquired it and just proceed to spend it. If, every time we faced a new context, however similar to this one, we had to pour into it the same quantity of labor, and labor of a new kind, we would be much less inclined to lose track of the first labor we put in and to diminish its dignity.

Having thus precisely located the difference between the two kinds of reasoning, we must avoid the temptation of (misleadingly) overstressing it. What allows us to prove the Euclidean theorem is one of infinitely many possible redescriptions of a triangle; once we hit upon it, the necessity of the conclusion strikes us with full

---

[3] "[C]onclusions which embrace more than we can grasp in a single intuition depend for their certainty on memory, and since memory is weak and unstable, it must be refreshed and strengthened through this continuous and repeated movement of thought. Say, for instance, in virtue of several operations, I have discovered the relation between the first and the second magnitude of a series, then the relation between the second and the third and the third and the fourth, and lastly the fourth and fifth: that does not necessarily enable me to see what the relation is between the first and the fifth, and I cannot deduce it from the relations I already know unless I remember all of them. That is why it is necessary that I run over them again and again in my mind until I can pass from the first to the last so quickly that memory is left with practically no role to play, and I seem to be intuiting the whole thing at once" (p. 38).

[4] "My propositions serve as elucidations in the following way: anyone who understands me eventually recognizes them as nonsensical, when he has used them—as steps—to climb up beyond them. (He must, so to speak, throw away the ladder after he has climbed up it)" (1961, p. 151).

[5] The role of a supplement, Derrida would say; see his (1974), pp. 141ff. For example, on p. 145, he writes: "But the supplement supplements. It adds only to replace. It intervenes or insinuates itself *in-the-place-of*; if it fills, it is as if one fills a void. If it represents and makes an image, it is by the anterior default of a presence. Compensatory and vicarious, the supplement is an adjunct, a subaltern instance which *takes-(the)-place*. As substitute, it is not simply added to the positivity of a presence, it produces no relief, its place is assigned in the structure by the mark of an emptiness. Somewhere, something can be filled up *of itself*, can accomplish itself, only by allowing itself to be filled through sign and proxy. The sign is always the supplement of the thing itself."

evidence. What allows us to prove the Hegelian point (for reasons given previously, I prefer not to call it a "conclusion") is one of infinitely many possible redescriptions of the master-slave relation; once we hit upon it, the necessity of the point strikes us with full evidence. Because of the formal character analytic logic owes to its disregard of most linguistic content, there is a tendency to concentrate on the formal conclusion thus reached and to disentangle it from the process whereby it was reached; however natural or even legitimate this tendency might be, it does not create more of a logical contrast between the two kinds of reasoning than already alleged and does not grant a nobler status to the necessity operative in one of them.

Now dismiss the heuristics and fasten on the logical relations themselves: on the strong sense that, when it comes to the magnitude of the angles of a Euclidean triangle, we are facing an unassailable truth, an effective and successful carving of (mathematical) reality at its joints, but, when it comes to the mutual dependency of master and slave, we are facing an exercise in rhetoric: an expert manipulation of words that suddenly makes it look as if things were the opposite of what we thought—much like the tricks played by Socrates in Aristophanes' *Clouds*.[6] To dispel this appearance of a sharp divergence, we need to work on both sides, and I begin with the geometrical proof.

First off, the proof only holds in *Euclidean* geometry. In its non-Euclidean variants, the conclusion we are discussing is not only not unassailable; it is not even a truth. Unassailably, there, something else is true: something contrary to what was proved above (the sum of the interior angles of a triangle is more, or less, than two right angles). And that is not all: it is not just Euclid's postulates that are optional; the same fate extends to his common notions. Without leaving analytic logic—that is, the Aristotelian-Fregean tradition—there is not a single logical principle (or

---

[6] And also *taught* by Socrates in that play (not quite effectively, as it turns out). Here is, for example, how poor Strepsiades practices what he has just learned: "**SOCRATES**: What would you do if a lawsuit was written up against you for five talents in damages? How would you go about having the case removed from the record? **STREPSIADES**: Er, I've no idea, let me have a think about it. **SOCRATES**: Be sure not to constrict your imagination by keeping your thoughts wrapped up. Let your mind fly through the air, but not too much. Think of your creativity as a beetle on a string, airborne, but connected, flying, but not too high. **STREPSIADES**: I've got it! A brilliant way of removing the lawsuit! You're going to love this one. **SOCRATES**: Tell me more. **STREPSIADES**: Have you seen those pretty, see-through stones that the healers sell? You know, the ones they use to start fires. **SOCRATES**: You mean glass. **STREPSIADES**: That's the stuff! If I had some glass, I could secretly position myself behind the bailiff as he writes up the case on his wax tablet. Then I could aim the sun rays at his docket and melt away the writing so there would be no record of my case! **SOCRATES**: Sweet charity! How 'ingenious.' **STREPSIADES**: Great! I've managed to erase a five-talent lawsuit. **SOCRATES** Come on, then, chew this one over. **STREPSIADES**: I'm ready. **SOCRATES**: You're in court, defending a suit, and it looks like you will surely lose. It's your turn to present your defense, and you have absolutely no witnesses. How would you effectively contest the case and, moreover, win the suit itself? **STREPSIADES**: Easy! **SOCRATES**: Let's hear it then. **STREPSIADES**: During the case for the prosecution, I would run off and hang myself! **SOCRATES**: What are you talking about? **STREPSIADES**: By all the gods, it's foolproof! How can anybody sue me when I'm dead? **SOCRATES**: This is preposterous! I've had just about enough of this! You'll get no more instruction from me" (pp. 53–55). No question about it: a lot of redescription is going on here.

rule), including modus ponens or that very epitome of rationality, the Principle of Non-Contradiction, without which (Aristotle believed; I will come back to this) we would not make any sense, that has not been criticized and denied, building competing logical systems on such denials.

Frege launched his entire logical adventure with the declared purpose of ruling out any reference to intuition from mathematical proofs.[7] And, in a way, he succeeded: whenever in the past a math teacher or speaker, in moving from one step to another of a demonstration on a board, would have appealed to his audience's common sense and license his inference by some such phrases as "from which there *obviously* follows..." or "it will be *evident* to everyone that...," now mention must be made (in principle, at least) of axioms or rules, or previous steps, that justify the move. In demanding that we make explicit the grounds of every logical inference, however trivial they might be, Frege unwittingly opened the way to the irrepressible efficiency of computers—which, notoriously, have low tolerance for what is generally understood and only react to literal formulations of the tasks involved. This is, by all means, an invaluable contribution on his part; but it is also an unintended one. Intuition he wanted to rule out from proofs not because it worked implicitly but because (by working *only* implicitly) it failed to justify the proofs. Intuition was taken by him as a gross form of wholesale agreement, and he thought that, by breaking it down into its constituent atoms, he could ground a proof on nothing other than the very laws of thought—which were implied, but not articulated, by such a wholesale nod. It was an unspoken premise of his project that intuition, when it worked, belonged to the same genus as the laws of thought and could be successfully resolved into them. Which, as later developments show, turned out to be well other than the case: fragmenting large steps into minute ones, useful as it is for unthinking machines bent on cranking up rigorous instructions, did nothing to expose the logical foundations of proofs. On the contrary, it exposed the fragility of logic itself.

Subscribing to a(n analytic) logical step, of whatever scope, requires two things: (a) a reference, however implicit, to the principles warranting the step and (b) a commitment to those principles. If the steps get smaller and smaller and

---

[7] Two significant passages among many: "Later developments ... have shown more and more clearly that in mathematics a mere moral conviction, supported by a mass of successful applications, is not good enough. Proof is now demanded of many things that formerly passed as self-evident" (1980, p. 1). "To this day, scarcely one single proof has ever been conducted on these lines; the mathematician rests content if every transition to a fresh judgment is self-evidently correct, without enquiring into the nature of this self-evidence, whether it is logical or intuitive. A single such step is often really a whole compendium, equivalent to several simple inferences, and into it there can still creep along with these some element from intuition. In proofs as we know them, progress is by jumps, which is why the variety of types of inference in mathematics appears to be so excessively rich; for the bigger the jump, the more diverse are the combinations it can represent of simple inferences with axioms derived from intuition. Often, nevertheless, the correctness of such a transition is immediately self-evident to us, without our ever becoming conscious of the subordinate steps condensed within it; whereupon, since it does not obviously conform to any of the recognized types of logical inference, we are prepared to accept its self-evidence forthwith as intuitive, and the conclusion itself as a synthetic truth—and this even when obviously it holds good of much more than merely what can be intuited" (ibid., pp. 102–103).

correspondingly the principles simpler and simpler, it also gets easier and easier to embed the principles into some automatic mechanism and "forget about it"; but it does not get easier to generate consensus on the principles—in fact, it perversely gets easier to find out how deep our lack of consensus can go and how elementary our disagreements can be. The multiplication of "logics" we have witnessed in the last few decades has nothing to do with the plurality of logics I am bringing out in this book (as I suggested in the preface): all of these "logics" fall within the range of one and the same general understanding of meaningful discourse; they are competing candidates for the privilege of accurately representing that general understanding (and often they are not even that: they are just little games of combinatorics played in order to get tenure). But it is important to realize that they are as much brainchildren of Frege's program as the modern computer (the Turing machine) is: on the positive side, the Fregean breaking down of proofs into minute steps favored the building of automata; on the negative side, it favored a calling in question of even the most basic logical components—those which, in the mist of "intuition," would have gone unheeded and would have been accepted as a matter of course.

And now for what relevance all of this has to us. If you voice with absolute Fregean forthcomingness all the most risible detail that goes into the identity of your formal system of geometry (or anything else), you can then forget about it and let your laptop run your "reasoning." Every step in the process that ensues will be so imposing for your mind that you might as well not have a mind: if you did, and assuming that it did not go to sleep or go mad in the presence of unassailable boredom, it would have to kneel in front of so much necessity. But if you change even the most risible detail of your formal system (thereby generating a different system), you might have to kiss that necessity goodbye and get ready to kneel in front of a different (a contrary) tyrannical imposition. The irresistible force the Euclidean proof exercises on your assent is (to repeat) a function of *two* things: the principles entering the proof, both geometrical and logical ones (both postulates and common notions), *and* your choice of committing yourself to those principles. If you turn that irresistible force into an absolute—as a premise to attempting to make the "relative" persuasiveness of Hegel's argument look bad—you are prey (or hoping to have others be prey) to an internal/external *Gestalt* shift: from within the system defined by the chosen principles, the force acts absolutely; but there is nothing absolute forcing the choice.

Move now to the other side of the coin. Each redescription of the master-slave relation will see it in a different light; and it will be left to our luck to find one that provides a fruitful view and left to our choice to select it and use it. But *if* we find it and select it—specifically, if we decide to look upon this relation under the guise of how the master depends for his mastery on the slave unilaterally, hence independently, accepting the role of a slave—how are we not constrained to follow Hegel's argument to the end, exactly as we were to follow Euclid's? How is one argument less forceful than the other: left hanging in the air of fleeting opinions rather than solidly anchored to the rock of objective reason?

I know how some will want to answer these questions, and addressing such answers will offer us another important angle on the issue. Euclidean geometry and

Fregean (or post-Fregean) logic are precisely identified structures: we know what their principles are and we can decide, in a finite time, if a given argument abides by those principles. We might want to go non-Euclidean in geometry or play by any of the myriad (analytic) logical games on the market today; but we will always know, precisely, what we are doing. With dialectical reasoning, no such precision is on hand: there is gesturing, there is enticement, there is (as I mentioned) rhetoric; you may come away persuaded, but you will never know (or be able to tell) *why*.

There is a good sociological basis for this complaint. After Frege (and Russell and Hilbert and Tarski) refined the analytic way of thinking and arguing, hundreds of their followers got busy exploring the potentialities of their proposals, pushing them and shoving them in all directions and changing them slightly or not so slightly, and in the process they (and we all) got clearer about what those proposals amounted to, freer and looser in exploiting them, more knowledgeable about the whole landscape. When we speak about precision gained, the cash value of what we are saying is to be found in this extensive manipulation of the tools of the trade, and in the consequent extensive comfort we developed in handling them, in the assurance we gained as we learned to walk on this terrain. As happens in every field, familiarity and confidence were acquired by playing, sometimes irreverently, with the steps made by previous masters and with the results they had obtained. With Hegel and other "continental" figures, the main attitude of their followers has been one of awe: of staring at their texts and never venturing to change a comma in them in fear that they might be causing irreparable damage, never trying an example of what is meant when a quotation will do. Some progress was made in this area, to be sure, and some variants were tried, but the reverence was always there, impeding simple matter-of-fact dealings, so it was as hard to detach genuine, original contributions from sheer repetition as was with many Medieval commentaries of Aristotle. If we feel more awkward walking around Hegel, therefore (as we did with Aristotle back then), it is largely because we have not done enough of it: his words have stunted (*we* have let his words stunt) the growth of our own.

Having admitted this weakness, we must, again, avoid overstressing it or misunderstanding it. The comfort and assurance we attain by getting used to a territory we often distill in tested recipes of behavior: strategies that have proved efficacious most of the time. And we might abbreviate those recipes by codifying them as formulas. The formulas, however, have only statistical value—they represent, I said, what works most of the time. They must not be confused with the formulas belonging to the systems I was discussing above: Euclidean or non-Euclidean geometries, Fregean or post-Fregean logics. There is a first internal/external confusion that threatens us, I noted: what is necessary inside a given system might not be necessary inside a different one. But there is also a second similar confusion possible between what happens inside *any* system and what happens within no system at all: in the unregulated, messy exchanges of everyday life. No matter how precise our systemic account of connectives is, we will have to face the fact that, as soon as we leave the rarefied atmosphere of abstract system building, even such a trivial connective as "and" will behave quite disparately in sentences like

(1) They are married and handsome.

(2) They had a child and got married

(while (1) is equivalent to

(3) They are handsome and married,

that is, (1) can be inferred from (3) and vice versa, (2) is *not* equivalent to

(4) They got married and had a child)

and hence our everyday recipes (or formulas, in one of *two* senses of this word) for how to deal with "and" will have to be taken with a grain of salt.

So, once more, we are comparing apples and oranges here. Whatever mileage Hegel's argument might have with us, the road it travels is an ordinary one, with all the bumps and fissures of ordinary conversations. If it convinces us up to a point, if it leaves us with the sense of something still unresolved, this is how it works in those conversations. The only generality that holds in them, I said, is statistical; the only normativity that rules the field is the frail and pliable one of our logical intuitions.[8] And because Hegel's logic cannot be formalized—that is, abstracted from context—there is no getting any better. Analytic logic, on the other hand, *can* be formalized: we can detach a logical point from its context and utilize it in all the infinitely many other contexts in which it is applicable. But what those contexts are will be decided by the same frail and pliable logical intuitions; *they*, and nothing else, will decide that, for example, the formal rules of and-introduction and and-elimination do or do not apply to statements like (1) and (2). There is a difference, then, in how necessity works in analytic and dialectical logic; but it is not that necessity works in one and does not in the other. It is (or, rather, part of it is) that the necessity at work in dialectical logic—the constraint[9] that is exercised by a piece of dialectical reasoning—works entirely at the level of ordinary language: there is no losing of any particle of meaning, no crystallizing of what is logically relevant in meaningless symbols, that is acceptable for it. In analytic logic, if we so wish, we can divide the

---

[8]The recurrence of the word "intuition" here, after the previous discussion of Frege, is significant and must be noted. For another way of phrasing the commitment to given principles that I considered crucial for subscribing to a step in a(an analytic) proof would be to say that one *feels* (or *has an intuition*) that the principles are right. Therefore, intuition *never* becomes irrelevant to the correctness of such a proof: whatever "laws of thought" we might reduce the proof to, they must be warranted for each of us by the normative intuitions (or feelings) I described in footnotes 5 and 7 in Chap. 2 and the attending text. This point is typically obscured by the maneuver I describe later: by dividing the issue of carrying a formal proof in an artificial calculus from the one of showing the relevance of that proof to ordinary contexts (and ordinary proofs) and then downplaying the significance of the second issue. Which explains, among other things, why the most difficult logic courses to teach are the most elementary ones, the ones in "informal logic," where logical intuitions are the very subject matter of the course and hence must be constantly referred to.

[9]Again, notice that a constraint is something we *feel* and, if we do not feel it, there is nothing that a formal system can do to make us feel it. But, again, in this case, it is a *normative* feeling (and constraint): a feeling that certain things *should* be done, and often will *not* be done, in a certain way—as opposed to the constraint that we feel, say, when we are chained to a wall.

problem (division continues to be its main strategy): conjure up a formal structure of supreme elegance in which consequences follow from premises with inexorable strength (to then have that structure fight it out with all of its alternatives, each as internally inexorable as it is externally optional) *and* use our ordinary, faulty, messy understanding of language to sort out the contexts to which those formal (and formidable) structures are or are not suitable. All well and good, provided we do not take the dividing to be conquering, that is, solving, the problem: provided we do not belittle the crucial *theoretical* significance of the second part of this division, which, however less elegant and less determinate than its companion might be, is just as needed to have the whole enterprise arrive at port. It is only if a large amount of such belittling makes us entirely forget such significance that we might incur the *illusion* at issue here (a logical illusion if there ever was one): that of contrasting inconclusive, somewhat irresponsible play with the ambiguities and fuzziness of language on the one hand and, on the other, an unrelenting, overwhelming weapon of mess destruction.

To summarize, analytic necessity fits the current ideological picture of how logical necessity works—how irrepressible and authoritative it is, how there is no alternative to submitting to it—only to the extent that we isolate it from the complexities of ordinary language and exile it within the confines of an artificial calculus, and that we keep at bay all the other calculi that, by their very existence, contest its would-be irrepressible verdicts. Independently of these disingenuous tactics, all there is to it is what we know already: the necessity by which contraries (however they might be identified by any one of us or within any of our systems) exclude one another. Move to dialectical logic now, and see how (else) necessity operates there.

I once read *Anna Karenina* with this concern in mind: if dialectical necessity is narrative necessity of sorts, I wanted to see how necessity worked in a true masterpiece of narration and learn from it. I noticed that, typically, major, even catastrophic developments occurred suddenly at the end of a chapter. In a brusque paragraph, Anna revealed her adulterous love to her husband or stepped in front of a train. What preceded such sharp turns of events was tens of pages of careful psychological analysis, in which nothing seemed to be going on but everything, in fact, was happening. What was happening was tectonic shifts orchestrated with infinite patience, minimal variations in the straws pressing upon a camel's back, subtle rearrangements of the reader's attention and emphasis, until what, earlier, would have seemed capricious and erratic ended up imposing itself—yes!—as eminently reasonable: surprising perhaps, but the kind of surprising that we experience when we see our hunches confirmed. If I were involved in literary criticism here, I might contrast this careful, *modern*, building up of expectations with the haphazard, evenemental scattering of events that is labeled *postmodern*, and I might point out that the latter is parasitical on the former: that the first globally resonant example of something haphazard taking place in a major work of narrative art—the shower scene in Hitchcock's *Psycho*—only had the powerful effect it did because of centuries of people being trained to expect otherwise and being given, as a result, the shock of their lives. But such is not the battle I am fighting here: what I am interested in is not the uttering of literary evaluations but the using of literary examples (modern, if that is what they are) to clarify a logical point.

Our competence as native speakers of a language gives us a feeling that articles and adverbs, say, can only be used in certain ways (that a sentence like "The does often" is not acceptable); it also gives us a feeling that, when a person, a situation, or an event is described in a certain way, certain other statements about the same subject matter must be coming. We call the first kind of feeling grammatical and the second logical, and I do not see much of a radical distinction there but that, too, is a controversy I leave aside now.[10] I concentrate on what is uncontroversially called logical and say that most of what makes one a great narrator is the capacity to deliver what Tolstoy does in *Anna Karenina*: describe a person, a situation, or an event so that our logical sense of what to expect next kicks in and the next thing we read is found by us to be perfectly in order. Or, rather, not just describe, for that would be tedious (how often do we hear it said that a novel, or a film, is a failure though it has splendid characters, because nothing happens there?), but *re*describe it: take something that appears well rooted and stable, stuck in a balance that might be excruciating but seems to offer no escape, and slowly throw a new angle on it, and guide it toward a *new* balance, which after a while might seem itself rooted and stable enough to grant a new perception of the thing, complete with lines of flight from the earlier prison. So here is what is the same and what is different in the two logics: The expectations we form as competent speakers based on how something is described in a novel or in a dialectical argument manifest the same necessity as the ones we form when following an analytic argument, and the shifting of these expectations as a consequence of redescription is a large part of how an argument (analytic *or* dialectical) makes that necessity relevant. The crucial difference is that dialectical logic then extends the necessity to the redescription itself, making it look as what it would have been rational to surmise, what we *should* have seen coming—and thereby making its necessity properly *narrative*: integrated in a course that, though twisted and torn, could not be otherwise, rather than dispersed in a number of individual steps, each individually necessary but not necessary *together*. Sooner or later, Anna's state of mind *had to* bring her to the point of walking away from her husband and in front of a train; sooner or later, the master and the slave must become aware of their co-dependency—and, if the story is told well, you will come to sense (after the fact, which is how stories work) such inevitability.[11] Therefore, if Galileo is right (as he is) and the best way of learning *analytic* logic is by practicing mathematical proofs, so as to get accustomed to the moves that are legitimate there, then

---

[10] I addressed some of it many years ago, within a debate with Gerald J. Massey, in my (1979).

[11] That the operation be conducted after the fact is crucial for conveying the sense of such inevitability; later in the text, I illustrate this point by telling a personal anecdote. Also, note that, by adopting a dialectical attitude, we could reconceptualize what is going on in the previous proof relative to Euclidean triangles as also demanding a particular redescription that makes our expectations shift and our sense of necessity evolve accordingly. Then the logic of the proof would incorporate its heuristics and necessity would pertain to the whole process, inclusive of the redescription. Which shows that, despite the prevalence of analytic logic in mathematics, dialectical logic may also be present there. I will show other elements of this presence in Chap. 9.

the best way of learning dialectical logic is by practicing (modern?) literary reading and writing. In enforcing mathematics to the exclusion of literary tales (except the most consistent, *mathematical* ones) in the education of his rulers (and, by implication, of his republic),[12] Plato was thus forbidding effective access to a whole way of thinking, reasoning, and arguing—as did all species of dictators forever since in their desperate attempts to stave off a change of the guard.[13]

I said that spacetime continuants are immediate cases of dialectical logic: obvious cases of identity in difference which are not, however, yet accounted for. Let us consider one such case and account for it in light of what was just explained. Take Socrates at five and (not to make it too hard on ourselves) Socrates at ten: many of their traits are different and yet, evidently, they are the same person. How are we supposed to redeem this immediate, inarticulate evidence? We will look away from everything that is simply different in the two kids: from the fact that Socrates at five is shorter, weighs less, and has more rounded features than Socrates at ten (I am making all this up, of course; "Socrates" here is just a name). We will focus instead on other elements that point to a uniformity between the two: both Socrates at five and Socrates at ten have the same sly countenance, the same mordant smile, the same impudent pose in front of grown-ups. We will argue that, whereas a variation in height and weight is not distinctive for kids of those two ages (they all mature pretty much the same way), it is his countenance and smile and pose that reveal Socrates' secret, that unfold his identity. *And* we will insist that this refocusing is necessary: that those who are not prepared to make it do not understand Socrates.

This is how dialectical logic transforms the opaque identity of continuants into a transparent one, immediacy into conceptual mediation: by applying to it the same narrative strategies we saw doing excellent work in a literary masterpiece. And what is true of the identity of an individual is also true of the identity of a community, of a nation, of humankind, or of the whole universe: if the challenge is rewriting chronology as argument, history as a well-connected story, it will be met by drawing attention at any crucial time, at any turning point, to some figures and features in what was until then the background and *fore*grounding them, making them serviceable to an appreciation of the crucial time as also being fatal, of the turning point as also implacably set, and thereby establishing that it is of *one* community, or nation,

---

[12] "'Shall we, then, casually allow our children to listen to any old stories, made up by just anyone, and to take into their minds views which, on the whole, contradict those we'll want them to have as adults?' 'No, we won't allow that at all.' 'So our first job, apparently, is to oversee the work of the story-writers, and to accept any good story they write, but reject the others. We'll let nurses and mothers tell their children the acceptable ones, and we'll have them devote themselves far more to using these stories to form their children's minds than they do using their hands to form their bodies'" (1993, 377a–c). "Left to ourselves, ... with benefit as our goal, we would employ harsher, less entertaining poets and story-tellers, to speak in the style of a good man and to keep in their stories to the principles we originally established as lawful, when our task was the education of our militia" (ibid., 398a–b).

[13] C. P. Snow's distinction of two cultures may have its basis in this logical difference. Though (as I mentioned in footnote 11 above) we will see that dialectical logic is not absent from even mathematics itself, it is certainly the case that analytic logic prevails in the ordinary practice of mathematicians and other scientists, and dialectical logic in the ordinary practice of humanists.

or humankind, or universe that we are talking—as opposed to a randomly collected pile of time-slices. As an additional example of how this operation is conducted in Hegel, consider the inception of the whole cosmic ride that his system consists of: Hegel's proof that the two most contrary concepts one could think of, those of *being* and *nothing*, amount to the same thing.

> *Being, pure being*, without any further determination. In its indeterminate immediacy it is equal only to itself.... [I]t has no diversity within itself nor any with a reference outwards.... It is pure indeterminateness and emptiness. There is *nothing* to be intuited in it.... Just as little is anything to be thought in it, or it is equally only this empty thinking. Being, the indeterminate immediate, is in fact *nothing*, and neither more nor less than *nothing*. *Nothing, pure nothing*: it is simply equality with itself, complete emptiness, absence of all determination and content.... Nothing is, therefore, the same determination, or rather absence of determination, and thus altogether the same as, pure *being*. (1990, p. 82)

If we focus on the opposition, the identity asserted here will appear absurd. But let us focus rather on what the content of *being* is and on what we are saying of something when all we say is that it *is*. What we are saying is totally empty; we are saying *nothing* of it, just as we would do if we said "instead" that it is not. *Being* and *nothing* are, equally, predicates of extreme, indeed total, indeterminacy; no matter how much we might *mean* by saying that something is, as opposed to saying that it is not, in both cases what we are actually saying is the same—we are saying nothing. And, if we do not see that, we are blind to what our very language is telling us, to what our very statements are urging us to realize—to the *necessity* expressed by that urge.

Because the issue is complex and delicate, one more example, not from Hegel but from everyday life, will not hurt. In my early 20s, I proved the independence of

(5) $\forall x \exists y (y = x)$

from the other axioms of a free logic with identity.[14] The way this happened, as I recall, is that one night I kept repeating (5) to myself out loud, until it hit me that a way to make this sentence false was by having the x's not necessarily coincide with the y's: by having different domains for different variables. If you think analytically, what I just said is the only possible account of what happened: I did a random thing and was hit by a random association. *After* receiving this random gift, logic went to work, the proof was initiated, and necessity took its course. If you think dialectically, on the other hand, then what I did was exactly the right thing to do, the thing that had to be done, and I had to do it repeatedly until the necessity of it, which was expressed by my own repeated words, made its way into my consciousness. If I had not done that right thing then, someone else (or even I myself) would have done it later; this is what had to happen anyway—it is the ineluctable logic of the situation. This example shows vividly (to me, at least, since I was there) that the two accounts are each a function of where you locate yourself with respect to the episode. It is *in retrospect*—that is, where dialectical logic, as we have seen, positions itself—that the whole thing feels perfectly natural, and you feel that it could not have gone any

---

[14] See my (1978).

other way; from within the episode, I am just mumbling to myself without a clue and being struck by a thought that comes from nowhere.

In summing up this discussion, I begin with what will never be said often enough: the outcome of this refocusing process cannot arrogantly dismiss the underbrush from which it materialized, because it is from that same underbrush that the next dialectical move, the next invocation of necessity, is expected, perhaps one that will contradict the one we have "established" now. What is in play here is the necessity of a life that never stops growing, and remaining attached to all of itself, as opposed to the necessity of a death that results when what is supposedly valuable is severed from everything else and contemplated forever in its unchanging, mummified state. But necessity it is nonetheless, in one case as in the other, the expression of a reason which, though differently modulated and implemented, espousing either the rigid repetition of the abstract or the continuous evolution of the relevant, advocating either scantiness or thoroughness, does what reason always does: overcome chance with order, shock with congruity, mystification with insight. Proclaim the *logos*: the surrender of chaos to meaning.

The necessity that operates in oceanic logic is different, in one fundamental respect, from both of those we have considered so far. Analytic and dialectical necessities, in their own distinctive ways, are brought into sharp relief by attention to detail: to the logical muscle formerly buried in grammatical fat or to the formerly imperceptible oddity or discrepancy that, once lighted on, will engineer a momentous rearrangement of the whole scenario. With our third contender, we are moving in the opposite direction: away from the fine print, toward a blurring of all meticulously drawn confines, all carefully established niches. You thought that pink was another color than red, or a bald man other than a hairy one, but you were wrong: everything is pink, to an extent, just as everything is also red; and every man is bald, just as he is also hairy. Diversity is only the coexistence of diverse points of view; but there is nothing definitive about the diversity of the latter—you can slowly shift among them and appreciate their various contributions to an understanding of the whole affair. You can see how there is a hairiness to the few stubs pushing forth from a perfectly shaved head or a baldness announcing itself in the parting of an otherwise unruly mane that shows a little too much scalp; so you can see how "hairy" and "bald" would be appropriate characterizations of what takes place in the one or the other case. (A beautiful woman looking at herself in the mirror and declaring herself to be horrible is not just voicing abysmal vanity: she is also making a logical point, too often unappreciated.) And, if you want to make others see it the same way, you will patiently disabuse them from the entrenched view that there is a destiny to the (analytic) configuration or the (dialectical) progress of anything: you will show them that everything, when looked at from the right (or—they will say at first—the wrong) angle, is exactly the same as everything else.

Necessity and identity go hand in hand: if X necessarily follows from Y, then X is, in some sense to be specified, *at one with Y*. It belongs to the same axiomatic system, say, or to the same spiritual progress—they are inseparable, implicated in each other. By centering on identity we can underscore another facet of the difference that is in question here. Take Socrates again and his identity across time. Ask

5 Necessity

yourself: *what* in Socrates (or *of* him) is identical across time? If you go analytic, it will have to be some version of Aristotelian essence: animality, rationality, or what have you. If you go dialectical, it will have to be some narrative pattern that cleverly overcomes oppositions by foregrounding consistency and continuity. Either way, what is identical will be intellectual and conceptual in nature: the concept of Socrates, perhaps, or the concept, period[15]—something that has nothing to do with the *matter* to which concepts apply, because in either view there is nothing identical in matter per se, in matter that is not conceptually qualified: you can never step into the same river twice, when the river is understood as the matter of it, as water. In oceanic logic, it is the other way around: *the matter* is what is identical. If all waves are modes of the same ocean, it is because they are all the same *stuff*, which irrelevantly takes such discordant shapes without losing its intimate concordance with itself. And it's the same godly stuff that can be described as omnipotent or eternal, as supremely good or just, as impassive or merciful; the same foodstuff that can be described as located at any point on a spectrum from sweet to bitter; and the same human-skin stuff that can be described as pink or brown. Within this logical framework, you not only can but *must* step into the same river all the time, and not even only that: you must step into the same body of water whether you step into a river or a lake, a puddle or a bathtub, here, there, and everywhere, because it is always (the same) waterstuff you step into, as opposed to the same analytically identified molecules of water, or dialectically unfolded water tales.

Since we have decided to resist the monistic tendency that drives oceanic (and dialectical) logic, and to treat it (indeed, to treat either) as a possibly local tool, we don't have to understand the *stuff* we are talking about (no matter how it was understood by the distinguished philosophers we referred to as committed to either logic) in any inflated metaphysical sense—as if we were speaking of the basic stuff of the universe, or if it even made sense to speak of any such basic stuff. All we need to refer to is some colloquial, unassuming distinction of matter and form, of subject and predicates, so that, when we deal with Socrates, we contrast the rough substance of his body with the qualities that from time to time, or forever, attach to that substance; and, when we deal with God, we contrast his countless perfections with whatever it is they are perfections of. With this humble contrast in mind, we will say that, when it comes to Socrates', or God's, or anything else's, identity, or to what is necessary to any of them, analytic and dialectical logic will search for it within the range of form and predicates, oceanic logic within the range of the subject of those predicates and the matter of that form. The first two will want solid ground to be discovered in what words evoke: a truer reality that condemns ordinary thickness as delusional, as a vanishing quantity, as shadows that can only fool prisoners in a cave. The third one will want to find it in that very brute, inarticulate thickness and will judge all articulation as colorful surface play: entertaining to be sure, but also distracting from the ultimate firmness of the real, and dangerous in insinuating the pretence of divisions in such firmness (all the perfections of Anselm's God are the

---

[15] There are no concept*s* in the plural in Hegel, just as there is no plurality of things: all the distinct concepts of the (analytic) tradition are but phases of the same intellectual structure.

same as God himself, remember), in turning the breaks suggested by the play into permanent rifts, "nature carved at its joints," or into ironclad chapters of an ineluctable narrative—all as a premise to possibly using this carving and storytelling for justifying the use of force, or (even more likely) to using force in order to keep the carving or storytelling afloat.

We have returned here, by another route, to the remark with which I closed Chap. 4. Oceanic logic, I said, offers intellectual dignity to the practice of compromise: what can only be understood by its rivals as a mediocre accommodation, for lack of anything better, or as a temporary, uneasy suspension of hostilities (if not as a ruse) is seen here as a perspicuous, penetrating appreciation of how mutually helpful different outlooks of the same thing can be, of how this is a single battle that all can win together (and can *only* win together, to the extent that the battle can be won). We can now see better why that is: what *must* be the case in oceanic logic—the necessity that is relevant to it—follows from the fact that we are always facing an identical thing. Pink must be brown not because the quality *pink* is the same (analytically or dialectically) as the quality *brown* but because *what is* pink is also *what is* brown, as shown by the uninterrupted path that, within that identical substance which is both, inevitably leads us from pink to brown. We can see this emphasis on matter, and this devaluing of verbal descriptions, as a stroke in favor of practice in general, not just the practice of compromise, and against theory, and to some extent it is. What is valued here is what most often goes unsaid—what we may even have no words for. But notice that all this emphasizing and devaluing happens *inside a theory*: it is a specific doctrine of the *logos* that so stresses the significance of what is *other than* the *logos*. It is a paradox we will have to take up again, when we have gone further comprehending this highly complex arrangement. For the moment, and for clarity, I bring this chapter to an end by showing the contrasting ways in which the three necessities we have discussed apply to one and the same case.

What is analytically necessary for Socrates has nothing to do with him, in fact he may be regarded as a distraction in acquiring proper knowledge of it: it is to be found in abstract patterns of which Socrates is but one of infinitely many instances and to which he is ultimately irrelevant except insofar as he (like infinitely many others, actual or possible) instantiates those patterns. What is dialectically necessary for Socrates has to do with Socrates only to the extent that his life can be rewritten as a thoroughly connected narrative and his substance can thus dissolve into the main character of that narrative; when this goal was attained, any demonstrative reference to Socrates would have been absorbed (and overcome) in a purely conceptual specification of spiritual development. What is oceanically necessary for Socrates has to do with his very substance: with the stuff that Socrates is made of—however ineffable such stuff might be—and of which his various qualities only provide ephemeral, fragmentary images, which show their fragmentary, ephemeral character (as well as the stuff's ineffability) by being constantly turned into each other.

Chapter 6
Truth

Abstract Analytic logic is obsessed with keeping reasoning free from error, as is seen most clearly in the developments of the philosophy of mathematics between Frege's work in the late nineteenth century and Gödel's second theorem: a sustained attempt at founding arithmetic on the bedrock of a logic that incorporated naïve set theory, the collapse of that attempt with the discovery of the paradoxes, Hilbert's program of vindicating all mathematical theories by proving their consistency, and a final resolution that no such proof was possible, even for the most elementary mathematical theory. Mathematics was then going to be practiced in the context of an ineliminable risk of error—and that was understood as a *limiting*, negative result.

In dialectical logic, error has no currency, because everything anyone believes, at any stage of the total narrative, is dialectically justified. There is still plenty of room, however, for relative error: for something that is false with respect to a truer, later phase of the narrative. And, in this sense, dialectical logic embraces error: the starting point of every dialectical argument is a rough, or immediate, position that through the course of the argument evolves into greater and greater truth.

In oceanic logic, truth inevitably becomes error: the ordinary truth that a man is uncontroversially bald, or white, when subjected to the relentless attacks of sorites, turns out not to be true after all—the man is just as much non-bald or nonwhite. This is seen most clearly in a class of paradoxes that are not often lumped together with the sorites but in fact express the same attitude: Zeno's paradoxes about space. There, too, uncontroversial truths about reaching a certain destination, or overcoming a contender in a race, are challenged and called in question. Ultimately, oceanic logic invites us to reject the arrogance of truth and learn to live in the territory of error.

The fundamental requirement on an inference is that it be truth-preserving: that it never lead from true premises to a false conclusion. That is what makes it a *good* inference. I am not going to inquire here into what truth is: to decide whether it be correspondence with reality (whatever *that* is), the coherence of a body of sentences, their pragmatic success, or anything else. I will remain neutral with respect to all such metaphysical views and only interested in the relation truth—however defined—has with the various doctrines of the *logos*. So what matters for me here, to begin with, is the fact that in analytic logic the main worry is to keep reasoning free of error. A valid argument is one whereby error does not spread: no new error (no error that was not already contained in the premises) is generated by it. If the premises of the argument are true (on which, however, logic most often will have no say), then error is kept entirely at bay: we can be sure that the soundness of our original theses is maintained through all logical steps and that the ocean of error will not contaminate the citadel of truth. It is not by chance that in the Fourth Meditation Descartes rehearses, while speaking of truth, the traditional defenses of God's providential goodness in the face of evil[1]: from the (traditional) point of view he incarnates there, error is *cognitive* evil, and it is just as mysterious how a benevolent God would allow his favorite creatures to incur in it as it is hard to understand how he would allow the suffering of innocent children.

Nothing documents the obsessive analytic concern with purity from this evil better than the half-century or so of turmoil in the philosophy of mathematics between the development of set theory by Frege and Cantor in the 1880s and 1890s and Gödel's (second) theorem in 1931. The first phase was heroic: arithmetic, perhaps all of mathematics, was to finally repose (as Odysseus did on the shore of Ithaca) on the bedrock of logic as such, to be recognized as just a more detailed, more complex form of thinking, reasoning, and arguing. Then error crept into this (apparently) safest environment: a sin that might be original was uncovered, a new exile from Eden was threatened, and a new archangel solemnly stated that this time it would not happen, that cognitive evil would not be as destructive as moral evil had been for that first earthly paradise—it would be exorcised.[2] The exorcism would have minimal ambitions: not error per se would be banned but only logical error, contradiction, and, for that goal to be attained, it was enough to prove that a single sentence was left out of any mathematical theory.[3] This most minute form of purity was going to be enough in order to vindicate the whole edifice of mathematical knowledge. And yet, even such a humblest of programs failed: as it turned out, it was going to be impossible to secure the area; people had to forever conduct their proofs under threat that, as in fact had happened to the founding father of the discipline (in its contemporary garb), the very ground on which these proofs were built might collapse under their feet.

---

[1] "As I reflect on these matters more attentively, it occurs to me first of all that it is no cause for surprise if I do not understand the reasons for some of God's actions; and there is no call to doubt his existence if I happen to find that there are other instances where I do not grasp why or how certain things were made by him. For since I now know that my own nature is very weak and limited, whereas the nature of God is immense, incomprehensible and infinite, I also know without more ado that he is capable of countless things whose causes are beyond my knowledge" (pp. 38–39). "I cannot … deny that there may in some way be more perfection in the universe as a whole because some of its parts are not immune from error, while others are immune, than there would be if all the parts were exactly alike" (pp. 42–43).

[2] "No one shall be able to drive us from the paradise that Cantor created for us" (Hilbert 1967, p. 376).

[3] "In the present situation … [the] problem of consistency is perfectly amenable to treatment. As we can immediately recognize, it reduces to the question of seeing that '$1 \neq 1$' cannot be obtained as an end formula from our axioms by the rules in force, hence that '$1 \neq 1$' is not a provable formula" (ibid., p. 383).

It would have been interesting to see researchers in the field interrogate the situation thus created: ask questions and propose articulate answers about what it means to do the most rigorous scientific work within the scope of essential undecidability. This did not happen: intellectuals of different brands and with no clear mathematical (or scientific) competence (e.g., deconstructionists) explored the issue in their own jargons and with their own conceptual tools, while mathematical logic and the philosophy of mathematics entered an extended sunset, or agony, dominated by routine, make-work, modest ambitions, and overall lack of enthusiasm. But the episode remains archetypal of the cosmic importance of error and of the need to stay clear of it: of how nothing serious can be comfortably entertained, within the kind of thinking made possible by analytic logic, if you believe that error might be looming, that at any time, unaccountably, it might rear its ugly head.

It is otherwise in dialectical logic. Here, generally speaking, no error is possible. The straightforward mistake of believing that $2 + 2 = 5$ has no currency, hence cannot be blamed as a corruption of rationality. What does have currency is the fact that a given person, in a given context, has come to believe that $2 + 2 = 5$; and there is nothing wrong with that—there could not be. Every sin can be forgiven, Hegel says, except the sin against the spirit; and that would be, precisely, the sin of claiming that something cannot be forgiven, that it is out of bounds for spirit.[4] Once the details of the occurrence of that idiosyncratic belief—or of something far less idiosyncratic but that we would still regard as an error, like the geocentric conception of the universe—were specified with complete accuracy, we would realize that it was exactly what had to be believed, by that person or by humankind, at that time and in those circumstances, that such is just what reason commands. Dialectical logic provides the ultimate theodicy, doing far better than Leibniz ever could with its analytic counterpart: whereas in the latter an appeal had to be made, in the end, to the incomprehensibility of God's plan in admitting evil, in the former evil is, simply, denied, recognized as the delusional outcome of poor vision, and dispelled when the light of reason shines on things and events and proves them to be perfectly adequate and suitable. Many of these justifications sound (I suggested already) as just-so stories, and some of them may strike one as plain offensive, as when Benedetto Croce, in his most Hegelian mode, rationalized the horrors of the first half of the twentieth century by mentioning the greater awareness of the value of liberty, and of the obligation to fight for it, that formerly complacent Europeans had thereby acquired[5];

---

[4] "[T]here is no sin that cannot be forgiven, except for the sin against the Holy Spirit, the denial of spirit itself; for spirit alone is the power that can itself sublate everything" (1984/1988 III, p. 235). "The sins of him who lies against the Holy Spirit cannot be forgiven. That lie is the refusal to be a universal, to be holy, that is to make Christ become divided, separated" (1995 I, p. 74; translation modified).

[5] In (2005), originally published in 1949, after he reports witnessing events that, as he puts it, "followed vertiginously, as if all the Furies, long kept repressed, having torn down the gates, had rushed into the world, igniting wars and revolutions, offering spectacles of horror that surpassed those of the barbaric ages, because they had available, in great plenty, the technical means provided by civilization," Croce offers as consolation the "pedagogical effect" these events had in exploding the "sense of security" people had acquired at the end of the XIX century, and in recalling everyone

nonetheless, such invariably optimistic tales are not (as I pointed out in Chap. 3) to be attributed to temperament (at least, temperament is philosophically irrelevant to them) but to the exigencies of the logic used. A logic in which error has no place.

On the other hand, if absolute error—error in and by itself—is ruled out, there is plenty of tolerance for relative error: for what is wrong, or false, with respect to something else that is more correct or truer. The most common employment of the word "truth," in dialectical logic, is not as a common noun labeling a quality of those particular things that are sentences (or maybe thoughts) but as part of an operator "the truth of" that applies to all sorts of things and refers (as, again, I mentioned in Chap. 3) to subsequent stages of their dialectical path.

You have a seed in your hands. Its structure is deceptively simple: you can't figure out from watching it, even examining it with the greatest care, what potential it has and what it can grow into—all of that is hidden from you and also (Hegel would say) from the seed itself. So you plant it in the ground, and water it and tend to it, and see it grow into a sapling and then into a large and beautiful tree, whose qualities are now quite apparent; as this gradual change takes place, you can say that the seed is coming more and more into its own, revealing more and more of its being, manifesting (to others as well as to itself) what of it was only implicit before. You can say that the sapling and then the tree are *truer* expressions of the seed or, in official Hegelian language, that the sapling is *the truth of* the seed and the tree the truth of the sapling (and of the seed—indeed *more* the truth of the seed than the sapling was). Just as Socrates at 70 is the truth of Socrates at five, or the full realization of a philosophical or scientific system is the truth of the intuition that first put its author on its track. And, if you want, you can travel the same road in the opposite direction and find in the seed a wrong*er*, fals*er* expression of the tree, one that is *more* in error as to what the tree is about—by coining neologisms that do not appear in Hegel's texts but are consistent with all that is found there (by irreverently playing with those texts as they would need us to do in order to reveal their vitality, and we would need to do in order to become familiar with them), you can say that the seed is *the error of*, or *the falsity of*, the tree (or the sapling).

Hegel has a hard time making a beginning of any of his works. His front materials (prefaces, introductions, and assorted other sections, in various combinations) take an inordinate amount of space and are often even more intricate and forbidding than his main texts.[6] Part of what makes him so hesitant to get started is that he

---

to the duty of "defending one's own faith and spending for it one's own forces" (pp. 301–302; my translation). A large part of what makes such complacent reconstructions possible, of course, is that they are *re*constructions: that, as we have seen, they situate themselves at the end of a given development and draw what is invariably a positive moral from it. As Kierkegaard illustrated in vivid detail, things look quite differently when you are situated inside that very development, with no clue as to how it will turn out; then, self-serving after-the-fact wisdom gives way to anxiety and turmoil. The personal episode I described in Chap. 5 brings out a similar dilemma.

[6] Of the major works published during his lifetime, the *Phenomenology of Spirit* contains a 45-page preface and an introduction; the *Science of Logic* contains two prefaces (for the two editions), an introduction, and a section of uncertain editorial role, since it nominally belongs to Book I but is titled *With What Must the Science Begin?* (some 60 pages altogether); the *Encyclopedia* contains

contests the very notion of a starting point, which is conceptually implicated with a linear progress (a beginning, a middle, and an end, as the philosopher would put it[7]), and believes the progress of spirit to be circular, with every point preceding and following every other point, explaining it and being explained by it. But there is another, related, reason for this awkwardness—one that brings us back to the matter currently at hand.

Given what I said above about truth and logic within the Aristotelian perspective, the main concern of an Aristotelian thinker or scientist is that of obtaining true principles: good inferences will then make it possible (indeed necessary) that truth never be lost and error never enter the proceedings. In order to arrive at those true principles which logic itself is impotent to deliver, Aristotle had to posit a separate intellectual virtue, *noûs*, which would allow its fortunate possessor to *see* what should be (but for many is not) evident: the cogency of the unprovable postulates without which a discipline cannot get started.[8] Coupled with *epistéme*, which made one recognize the cogency of logical steps in a proof, *noûs* constituted *sophía*[9]: the state of mind one needed to practice (good) Aristotelian science. Non-Euclidean geometries dealt a lethal blow to the notion of some statement being evidently true and opened the way to a different sort of uncertainty than the one later unleashed by Gödel: as Plato's *Republic* had already warned us,[10] the principles of a science (for

---

three prefaces (for the three editions), an introduction, and (again!) a section that nominally belongs to the First Part but is in fact a Preliminary Conception of it and summarizes the entire previous history of philosophy, followed by a More Precise Conception and Division of the *Logic*, for a total of over 130 pages; the *Elements of the Philosophy of Right* have a preface and an introduction that together come to over 50 pages.

[7] See Aristotle's *Poetics* 1450b.

[8] "Knowledge is belief about things that are universal and necessary, and there are principles of everything that is demonstrated and of all knowledge (for knowledge involves reasoning). This being so, the first principle of what is known cannot be an object of knowledge, of art, or of practical wisdom; for that which can be known can be demonstrated, and art and practical wisdom deal with things that can be otherwise. Nor are these first principles the objects of wisdom, for it is a mark of the wise man to have *demonstration* about some things. If, then, the states by which we have truth and are never deceived about things that cannot—or can—be otherwise are knowledge, practical wisdom, philosophic wisdom, and comprehension [*noûs*], and it cannot be any of the three (i.e. practical wisdom, scientific knowledge, or philosophic wisdom), the remaining alternative is that it is comprehension that grasps the first principles" (*Nicomachean Ethics* 1140b–1141a).

[9] "Knowledge (*epistéme*), then, is a state of capacity to demonstrate and has the other limiting characteristics which we specify in the *Analytics* [where the theory of syllogisms is spelled out]" (ibid. 1139b). "[T]he wise man must not only know what follows from the first principles, but must also possess truth about the first principles. Therefore wisdom [*sophía*] must be comprehension combined with knowledge" (ibid. 1141a).

[10] "[D]ialectic is the only field of enquiry which sets out methodically to grasp the reality of any and every thing. All the other areas of expertise, on the other hand, are either concerned with fulfilling people's beliefs and desires, or are directed toward generation and manufacture or looking after things while they're being generated and manufactured. Even any that are left—geometry and so on, which we were saying do grasp reality to some extent—are evidently dreaming about reality. There's no chance of their having a conscious glimpse of reality as long as they refuse to disturb the things they take for granted and remain incapable of explaining them. For if your starting-point is unknown, and your end-point and intermediate stages are woven together out of unknown mate-

Plato it was a mathematical science) are but assumptions and hence are arbitrary. As such, they can hardly provide an epistemic foundation for their consequences; if anything, some iteration of the hypothetico-deductive model will have the consequences epistemically justify the principles, insofar as those consequences can be factually tested and confirmed. In the new picture, the virtue of *noûs* is not so much dropped as it is reinterpreted: it still takes a certain mind (say, Einstein's) to come up with powerful and fruitful assumptions that will later be confirmed by a vast array of facts; and it is still a combination of that virtue with the one displayed in keeping track of what follows from what that makes for a good, or great, scientist. But the ideal of a supremely robust edifice built on sound, unshakeable ground and erected with equal soundness one story at the time is going to remain just that—an ideal. The proud confidence intimated by Aristotle and (apparently) vindicated by Euclid is, by now, a thing of the past.

Be it as it may with analytic ideals and reality, dialectical logic is situated otherwise. An argument here will develop more truth out of less—if you will, truth out of error. What was formerly primitive, inchoate, and mute will become eloquent, articulate, and intelligible, without of course changing into something else—indeed bringing out more and more of what it always was, more and more of its truth. This means that the meticulous search for the true, most general principles from which every other truth should follow gives way here to total indeterminacy: we can start wherever we please, and the forceful internal logic of that arbitrarily chosen content will take us everywhere we want to go—which, for Hegel, is *everywhere*. There is nothing special about any particular beginning, which is the second reason why, if you write a Hegelian text, you have a hard time justifying the subject matter of your first page: you could replace it with what comes ten or a hundred pages later and you would be writing (dialectically) the same text, eventually circling back (to return to the first reason discussed above) to what you now put on page one.

This whole rumination about beginnings is a pointer for the crucial issue to be emphasized here: if in dialectical logic you must start with error to arrive at truth, it is because truth is inextricably linked with error. Rather than a stable configuration to be calmly appreciated with the eyes of the mind (in analogy with that sight which, of all senses, is for Aristotle the most conducive to knowledge[11]—being the one that least impacts the object sensed), truth is here (as suggested by the operator which is its main grammatical context of occurrence) a process of having something overcome a previous state and reveal more of its dynamical identity: there is no truth in quietude but only in friction and unrest. The destiny of error is to turn into truth without losing itself, indeed taking what was errant in it into a new phase and

---

rial, there may be coherence, but knowledge is completely out of the question" (*Republic* 533b–c).

[11] "All men by nature desire to know. An indication of this is the delight we take in our senses; for even apart from their usefulness, they are loved for themselves; and above all others the sense of sight. For not only with a view to action, but even when we are not going to do anything, we prefer sight to almost everything else. The reason is that this, most of all the senses, makes us know and brings to light many differences between things" (*Metaphysics* 980b).

making it, when seen from the appropriate, revolutionary point of view, the basis of its attaining truth. It is the caducity of Socrates, his being composed of feeble, mortal spoils—a limited, rudimentary, *mistaken* view of him—that provides the substance and meaning of his immortality, which is (as I said before) the immortality of a human, *the immortality of a mortal being*.

When it comes to the confrontation of logics, every parameter and every value are at stake; so we can also take another look at the purity that was seen to characterize the analytic relation with error. I called it "obsessive," and the term is not unfitting, since the attitude in question reminds one of those who would not touch doorknobs or shake hands for fear of being infected by deadly viruses. Plato's judges in the *Republic* are supposed to be pure like that (just as his citizens, I noted above, are supposed to be pure of all artistic diversity): since their souls are the tools by which they judge the crimes of others, they must not be tainted by any direct experience of crime; only late in life, and only by indirect routes, must they pick up a minimal amount of knowledge (by description, not by acquaintance) of the deplorable behaviors on which they are to pass judgment.[12] Such is analytic purity; but it is not the only kind. There is another mode of purity, which is appropriate to dialectical logic. A person who is pure in this mode is someone who has been through all walks of life, high and low, clean and filthy, who has navigated the restful waters of temperance and the stormy ones of addiction and sin and has made contact with the whole range of what these diverse encounters can bring out in herself, and who has not so much safely maintained as laboriously gained and regained her integrity, and can look at the entire course of her life without forgetting or regretting anything (though possibly being ashamed of something), feeling stronger for all that has not killed her (in case you wonder, Nietzche got a lot from Hegel, though with little mention of him). Once again, this purity is not a *fait accompli* but a constant struggle: it is not the purity of those who stay away from deadly viruses but the one of those who inoculate themselves with them, being always out there and being always conscious that there are more viruses out there than the ones they have already learned how to handle.[13]

Similarly, someone who addresses knowledge by dialectical logic will not shut herself up inside the prison of academia, of illustrious references and weighty

---

[12] "[I]t's out of the question for a legal expert to be brought up, from childhood onwards, in the company of minds which are in a bad condition, and for his mind to have thoroughly explored the whole arena of immorality until it has become immoral itself and can quickly use itself as a criterion by which to assess the immorality of others' actions.... Instead, his mind must, while young, have no experience of bad characters, and must not be contaminated by them, if it's to become truly good at assessing the morality of actions in a reliable manner.... A good legal expert must have been slow to learn the nature of immorality, because he's been observing something which is not an inherent quality in his own mind, but an alien quality in other people's minds. He must have trained himself over many years to discern its badness by making use of information, not his own experience" (409a–c).

[13] This is not just a metaphor (is anything?): choosing immunization requires adopting a certain logic of "health," and so does refusing vaccines. Similar remarks apply to, say, displaying your valuables versus keeping them under lock and key. (Venice has been compared to a theater and Genoa to a safe.)

quotes, writing "essays" that are nothing but: that start with established, authoritative statements and add one or two of the same. She will interrogate the errancy of knights young and old—astrology and the Internet, texting and literature, alchemy and chemistry—and from each site of error will spin a thread that transfigures it, that connects it with infinite other details and by each connection reveals more of its destiny, fulfills more of its promise, manifests more of its surprising scope, of the surprisingly deep contribution it makes to global comprehension. This will be the purity of a public intellectual who is not afraid of really *essaying* (of trying, that is): of entering any debate and learning from it while not selling out. On the contrary, she will provide some dignity to what might have seemed undignified and spell out the conceptual significance of what might have seemed extraneous to the concept. (I will return to this theme in Chap. 11.)

To summarize, in analytic logic truth stays truth and error stays error, whereas in dialectical logic error inevitably becomes truth. In the same crisp terms, in oceanic logic truth inevitably becomes error. You start out thinking that this man is white, period, or is just as uncontroversially bald: such is the truth of the matter, part of the correct carving of the man's existential space. Then oceanic logic goes to work and shows you that the man is just as black as he is white, just as hairy as he is bald, and that he is white, or black, or bald, or hairy, or anything else, only in a manner of speaking, only when he is regarded from a certain point of view, and certain features of his are stressed to the detriment of others and as a result he is, indeed, *spoken* about in a certain *manner*. That all different points of view are useful and legitimate, even when they entail opposite conclusions, means that none of them attains the true essence of the man, what the man really is (as opposed to anything else—whether "opposed" is understood analytically *or* dialectically); once again, this logic announces the limitations of the *logos* itself, its being reduced to a dance over the surface of reality—an engaging, delightful, even profitable dance, but one where feet never touch the ground.

The most extreme form of this attack on the conventional understanding of how language relates to reality oceanic logic conducts by another class of arguments, close in structure to sorites but typically not lumped with them. They do not come from Megara but from Elea: their goal is not the polemical one of countering Aristotle but the positive one of sustaining (by polemical tools) a deviant metaphysical outlook, Parmenides'.[14] They are Zeno's paradoxes, to some of which I turn now.

You are 10 ft away from a desired object. You have complete control of your limbs and motion. You walk toward your goal. There is no question that you will get there in a few seconds and experience the joy of possession and fruition. Or is there? The goal you have set for yourself is on this side of the split between the contraries

---

[14] Without entering here into complex and controversial interpretative issues, I can at least point out that the most general understanding of Parmenides' only work, the (fragmentary) poem *On Nature* (for which see Kirk et al. 1983), is germane to oceanic logic as I am presenting it here: it contrasts a world of reality that is whole, uniform, timeless, necessary, and continuous with a world of opinions and appearances that induces separation and change.

reachable and *unreachable*: a clear case of the former. If you had set as a goal something to be found at the end of the world, or maybe just in the next galaxy, that would be a clear countercase of it—and a clear case of *unreachable*. But there are, here as on previous occasions, cases in between, and they confirm their perplexing tendency to infect all others. As long as there is space separating you from your objective, that space can itself be divided indefinitely, making for indefinitely many distinct moves traversing each segment. The first move traversing the first infinitesimal segment does not take you where you want to go and, if you did not get there so far, yet another move traversing another infinitesimal segment will not make a difference; so you will never get there. The reachable has become unreachable.[15]

I will address the problems raised by such examples in two successive phases. First, any number of things could happen within 10 ft. As the move that traverses them splinters into an uncontrollable mass of mini-moves, the choice I originally made to perform that move (globally understood) also splinters into an uncontrollable mass of mini-choices, all equally and jointly required to attain the final outcome. That I made any one of them does not force me to make any subsequent one; I could stall at any point and no longer care to go forward. Hannibal left North Africa and traversed Spain, the Pyrenees, the South of France, the Alps, and most of the Italian peninsula only to stop a few miles from his supposed destination and sit there for some 15 years, until he turned around and went back to North Africa. An action is made of indefinitely many actions and is not completed until they are all completed. So, *before* that moment, the completion is going to look impossible, as you consider how much must happen (and you must do) exactly right in order to give it a chance. You could divide up the task into a few sizable chunks and only recognize (and mark) the time when a particular chunk has come to an end; but oceanic logic will insist that this strategy is disingenuous—that there is no limit to how many chunks could be recognized and to how many marks could be made. And, therefore, that whether a task requires finite effort, finite time, and finite energy or the opposite depends on how you look at it and on how you decide to describe it. What Hannibal did was what Hannibal did: he went up this mountain and across that lake; he slept in tents and raised his sword; he caught the Romans by the balls at Cannae and enjoyed himself at Capua. Whether all of that was part and parcel of *getting to Rome and conquering it* depends on how you talk about it; and, if you talk about it that way, then of course it looks as if what was supposed to be the last step

---

[15] The reason why these paradoxes are typically not lumped together with the sorites, I believe, is that superficially they seem to be going in opposite directions: the latter make us merge together what appeared radically divided; the former make us irreparably divided from what appeared to be in our grasp. But this superficial impression can be reversed (thus helping us appreciate the deep similarity between them): you thought that you could go smoothly from a non-heap to a heap and yet, one grain of sand at a time, you will never get there; you thought there was a radical distinction between what you can and what you cannot reach, and yet there isn't—the reachable is also unreachable. In fact, as will surface shortly in the text, both sorts of oddities originate from the same source: there is an irreducible heterogeneity between the continuity of the world and the discreteness of the measuring—or conceptual—devices we superimpose on it. A heap is not made of grains of sand, and a distance is not made of individual, separate points.

of his itinerary took him forever, and completion never came. Similarly, whether you will make it to your declared goal 10 ft away remains to be seen.

But now suppose I do *not* stop. Do I not, then (and he), reach the destination? And what about Achilles not so much reaching as overcoming the tortoise? Isn't that what we expect to take place all the time; what *does* take place all the time?

Not so fast (!). To begin the second phase of my remarks, let us reflect on a colorful anecdote. A tale is told that, when Diogenes was presented with a like argument to the effect that movement does not exist, he stood up and moved. Which, for many, is what is paradoxical about this whole line of thinking, what qualifies it as literally "beyond belief." For movement *is*, hence an argument that seems to show that nothing moves is incredible, absurd, and must be rejected. But deeper inspection shows that what Diogenes did is irrelevant to the argument, which does not pertain to what is but, once again, to what we *call* what is, hence also to how we *understand* it. Diogenes stood up from being seated, and we may call that moving if we so wish; but we may also call it being still on the same spot (in two different modes); and if he took a few steps we may call that being still in the same location, and if he traveled to Sicily we may call that being still in the same continent. There is nothing Diogenes can do that forces us to avoid the terminology of stillness in describing his behavior; movement and stillness are manners of speaking, and depending on how we speak of him Diogenes will be one or the other. (Is Diogenes moving when he just sits? He breathes, doesn't he? Do we call that moving, or not?) In the same way, we may say that the Magdeburg hemispheres were empty in an analytic sense: that they contained a vacuum, a contrary to what is occupied. Or we may say that being and nothing are empty concepts in a dialectical sense, where their emptiness is the prologue to immeasurable riches: it is implicit fullness. *Or* we may say, oceanically, and as both Descartes and Sartre have said, that a room (or cup or coffee shop) is empty, when in fact it is full of things and people, because we choose to look at it from the point of view of what we were interested in finding there, which is not to be seen.[16] Within this logic, what is there is always the same, and it may be empty or nonempty depending on the language we use.

What I said about Diogenes can be generalized. There is no denying that I can get very close to my destination, even infinitely close to it; does that amount to reaching it, touching it—is that what we should call what I did, how we should understand it?

---

[16] "In its ordinary use the term 'empty' usually refers not to a place or space in which there is absolutely nothing at all, but simply to a place in which there is none of the things that we think ought to be there. Thus a pitcher made to hold water is called 'empty' when it is simply full of air; a fishpond is called 'empty,' despite all the water in it, if it contains no fish; and a merchant ship is called 'empty' if it loaded only with sand ballast. And similarly a space is called 'empty' if it contains nothing perceivable by the senses, despite the fact that it is full of created, self-subsistent matter; for normally the only things we give any thought to are those which are detected by our senses" (*Principles of Philosophy*, p. 230). "I have an appointment with Pierre at four o'clock.... But now Pierre is not there.... So that what is offered to intuition is a flickering of nothingness; it is the nothingness of the ground, the nihilation of which summons and demands the appearance of the figure, and it is the figure—the nothingness which slips as a *nothing* to the surface of the ground" (Sartre 1992, pp. 40–42).

Maybe so but, also, maybe not. It all depends on what we take reaching and touching to be, on how we think of them. Heidegger—not coincidentally, an author we saw committed to oceanic logic—claims that physical bodies never touch.[17] What forever remains between them, however imperceptibly, is a reminder of what Zeno might have been getting at (without touching it?). As we dig deeper and deeper into the physical structure of matter—the one studied by mathematical physics, the one that calculates the 10 ft between me and the object—we always find parts outside parts, parts separated from each other by space. Perhaps a space that reduces to nothing, but still signals the externality of one thing to the other, their not becoming *one*, and hence also a space that stays unconquered. If I ever do get to the thing, if I ever succeed in reaching and touching it, it must be by some other means than getting there one step at a time. Perhaps I was touching it even before I left. Perhaps reaching and touching have nothing to do with physical distance.[18]

As for overcoming, it makes sense to talk about it when two things are traveling on the same line, and eventually one of them makes contact with the other and then goes past it. There is overcoming in dialectical logic, when Socrates makes contact with a stage of his development (how could he not, since that stage is *him*?) and then transcends it into a new stage. But that is not how it goes with Achilles and the tortoise. If Achilles never makes contact with the tortoise, as the previous point was intending to prove, then they simply travel in different directions; and how can one overcome the other, then?[19] Am I overcome by another car when I am driving, if that car is going somewhere else? (Try telling that to those who roar by your side window, wanting to make you feel humiliated and sad but in fact just going their own way. If, that is, you can make contact with them.)

And now for the significance of all of this. That I will not reach a goal 10 ft away from me, that the arrow will not hit its target, that Achilles will not overcome the tortoise are ways of saying that measurement does not match the world: it does not resonate and agree with its intimate structure. It is, rather, superimposed on a recalcitrant world, one that resists being firmly labeled, a horse that repeatedly throws off its rider as he attempts to tame it. The net we cast on reality to take stock of it only takes stock of itself. We do not evaluate what separates two things by lining a yardstick between them; we do not have one thing "finally" reach the other when the yardstick has been successfully retired. Reaching something means merging with it, overcoming its outsideness, and establishing continuity with it—none of which can be defined in terms of inches and feet. And what is true of measurement

---

[17] "[I]n principle the chair can never touch the wall, even if the space between them should be equal to zero. If the chair could touch the wall, this would presuppose that the wall is the sort of thing 'for' which a chair would be *encounterable*" (1962, p. 81).

[18] In Chap. 4 I showed that sorites are reversible: by reversing the one proving that every man is white, you could prove that every man is black. The same is true of Zeno's paradoxes (which is another point of resemblance between them). Assuming that I am touching something, would moving away from it by an imperceptible, infinitesimal degree amount to *no longer touching it*?

[19] Mathematical physics will tell you that Achilles' movement can be divided in such a way that one component of it is on the exact same line as the tortoise's. But that, as will become clear below, is one more case of superimposing a measuring/conceptual scheme onto a world that is foreign to it.

is true of conceptual specifications in general: they come and go; they contradict each other; and, in doing so, they contest the very idea that they could express the truth of the situation. Truth is the result of narrow vision: of being stuck with one particular angle and not even imagining what other angles could reveal. Of sitting here, seeing the food, the gold, or the woman, 10 ft away, and quickly calculating that we will be there in a few seconds. But oceanic logic argues that there are countless barriers to be crossed, as many as there would be in a 1000 mile trail; and then perhaps we never stand up and begin our journey. Or we do, as Diogenes did, and we even "get" there; but that, as I said, proves nothing. Because, possibly, there is no "there" that we can get to—no prize that we can grab—that way: what we grab that way might never be the prize we were aiming at. (And you should listen when the woman says that you have not reached her, for all your effort: that you are sorely missing her; that you, in fact, have been avoiding her. She is giving you a logical lesson, among others.)

Bergson deals with Zeno's paradoxes and makes two points about them which parallel the two I just made. Taking them in reverse order, we have, first, a "cinematographical" way of reading the world that does constant violence to it:

> Suppose we wish to portray on a screen a living picture, such as the marching past of a regiment. There is one way in which it might first occur to us to do it. That would be to cut out jointed figures representing the soldiers, to give to each of them the movement of marching, a movement varying from individual to individual although common to the human species, and to throw the whole on the screen. We should need to spend on this little game an enormous amount of work, and even then we should obtain but a very poor result: how could it, at its best, reproduce the suppleness and variety of life? Now, there is another way of proceeding, more easy and at the same time more effective. It is to take a series of snapshots of the passing regiment and to throw these instantaneous views on the screen, so that they replace each other very rapidly. This is what the cinematograph does.... Such is the contrivance of the cinematograph. And such is also that of our knowledge. Instead of attaching ourselves to the inner becoming of things, we place ourselves outside them in order to recompose their becoming artificially. We take snapshots, as it were, of the passing reality, and, as these are characteristic of the reality, we have only to string them on a becoming, abstract, uniform and invisible, situated at the back of the apparatus of knowledge, in order to imitate what there is that is characteristic in this becoming itself. (1911c, pp. 321–323)

Here cinematography is the measuring, or conceptual, stick we bring to reality, which ends up giving us only the illusion of reaching it, while it reaches only itself. But why do we do this? Because our interest is practical: we are not so much interested in what reality is as in what we can do with it—not in what the distance really is between us and an object but in how the object can be manipulated and made serviceable to our purposes.

> The function of the intellect is to preside over actions. Now, in action, it is the result that interests us; the means matter little provided the end is attained.... The intellect ... only represents to the activity ends to attain, that is to say, points of rest. And, from one end attained to another end attained, from one rest to another rest, our activity is carried by a series of leaps, during which our consciousness is turned away as much as possible from the movement going on, to regard only the anticipated image of the movement accomplished.... Our activity is fitted into the material world. If matter appeared to us as a perpetual flowing, we should assign no termination to any of our actions.... In order that our activity may leap

> from an *act* to an *act*, it is necessary that matter should pass from a *state* to a *state*, for it is only into a state of the material world that action can fit a result, so as to be accomplished. (ibid., pp. 315–316)

And yet, we should acknowledge that our moves are as fluid and uninterrupted as the world is, hence also equally recalcitrant to firm divisions, equally alien to falling neatly into a predetermined place. We should ponder over the fact that even someone as willful and deliberate as Hannibal stopped short of his intended goal. We should realize that the effort of reaching out to things might prove vain; that there may be altogether different ways of conceiving what we are doing.

Bergson judges Zeno's arguments to rest on an illusion:

> We ... obtain a series of absurdities that all express the same fundamental absurdity.... The absurdity vanishes as soon as we adopt by thought the continuity of the real movement, a continuity of which every one of us is conscious whenever he lifts an arm or advances a step. We feel then indeed that the line passed over between two stops is described with a single indivisible stroke, and that we seek in vain to practice on the movement, which traces the line, divisions corresponding, each to each, with the divisions arbitrarily chosen of the line once it has been traced. The line traversed by the moving body lends itself to any kind of division, because it has no internal organization. (ibid., p. 327)

Aside from the usual arrogance of professional philosophers, this firm conviction of having resolved, once and for all, a problem that goes back millennia is what we would expect to follow from the commitment to a logic—as are the accusations of absurdity launched at those who think otherwise. If you think oceanically, then continuity (a word we have seen Bergson use before[20]) comes naturally to you, and attempts at capturing it with discrete units of measurement look hopeless. But, for those who do think otherwise, a commitment to those units may come just as naturally: what you take to be wrong may be for them an inevitable element of how sense is made of the world, which, by a rather circuitous route, takes us back to the topic of truth and error and offers a useful way of bringing the present chapter to a close.

We are used to thinking that an error is just a mistake: something to be avoided and corrected. That, it turns out, is one more consequence of our entrenched allegiance to analytic logic, for which to err, to wander, to roam about is a source of distrust and disapproval: for which matters must stay put, resign themselves to forever inhabit the trenches (or is it tombs?) we have dug for them as we were busy carving nature at its joints. From the rivals of analytic logic we hear two different sorts of protestations against this negative attitude: two distinct ways of conceiving of the "joints" that are "carved" as anything but natural—as the result of a repression, as needing to be maintained by barbed wire. We are provided two distinct strategies for redeeming error, and erring. One incorporates the vagaries of error into truth itself, since truth is in motion, constantly aiming at a destination that transcends its current state, and wandering is the form that motion takes. The other

---

[20] In the passages from (1911b) and (1911c) cited on p. 48. In Chap. 9 we will see the contrast between continuity and discreteness emerge again, as we discuss the presence of oceanic logic within mathematics.

refuses to even consider a destination, or a destiny, and keeps error as an impertinent, relentless irritant for anything that at any time would be regarded as truth. This kind of regard is now blamed as a deadly stare, which must be countered by throwing a new, unexpected angle onto the proceedings: one that makes the previous banality suddenly look paradoxical.

As we appreciate the fact that this (latter) provocation presumes to bring out a new perspective *on the very same thing*, we might think of it as expanding the range of what it means to be true. God is not omnipotent *as opposed to* omniscient or eternal; his omnipotence is but another description of his omniscience, or eternity—and, when we understand that, we understand God better: we get more of the truth about God. Except that we get it, once again, not so much within as beyond language. God is something unspeakable that we might call omnipotent, or omniscient, or whatever: the most important result attained by the errancy among these names is to disestablish the claim any of them might make to be what, in language, faithfully represents that unspeakable reality. (Which is what negative theology has been trying to tell us all along: of God, we can only say what he is not, thus only expressing the limits of our speech. Oceanic logic generalizes this attitude to everything other than God—though in the next chapter we will see that within oceanic logic itself, as opposed to its polemical use against its analytic rival, "negative" here should be replaced with "differential.") We can call a reference to this reality "truth" if we so wish, as long as it is clear that it is not a truth we can access or voice. It is a truth we have no grasp of; on the other hand, any truth we might have considered accessible or expressible is now given the lie by being shown to be no other than one of countless points of view, each detailing the structure and needs of the viewer rather than the objectivity of the viewed. For all practical purposes, truth has indeed, as I promised, been reduced to error. And the attitudes on truth and error of the three theories of the *logos* can be summarized in the following three injunctions:

(a) Stay within truth, and away from error.
(b) Do not be fooled by the contrast between truth and error: develop the truth contained in error.
(c) Reject all claims to truth; learn to live forever in the territory of error.

Chapter 7
Negation

Abstract The most characteristic word of analytic logic is the adverb "not"; everything in this logic is defined by what it excludes; and the Principle of Non-Contradiction (which might be better called Principle of Non-Contrariety) is the logic's most basic dogma. Therefore, proofs by *reductio ad absurdum* (which proceed by reducing a thesis to a contradiction) are here a basic tool, and a question arises (which I put aside here) of how far such a proof can be turned into the proof of a positive conclusion—that is, how far one should trust the Principle of the Excluded Middle.

In dialectical logic, negation does not exclude; indeed, this logic thrives on overcoming negation. But there is a large issue in Hegelian scholarship concerning whether, when something turns into its own contradictory, what contradiction there was is to be understood as retained or as canceled. The answer to this question is that it is both: Hegel's crucial word "*aufheben*" means both. The two contradictory items are both retained in the new phase of the narrative; but they themselves have evolved in ways that allow for a reconciliation between them.

In oceanic logic, exclusion plays no role and negation, being its linguistic expression, plays none either. It is hard to come by this realization because negation often occurs in the formulation of sorites, which are typical oceanic arguments; but sorites are weapons to be used against an enemy—analytic logic—so it is no wonder that they be phrased so as to bring out the difficulties of analytic logic, in *its* language. In proper oceanic language, there is no negation but only difference: the different points of view that can be taken on one and the same thing and that only ever capture different, but compatible, aspects of it.

The Aristotelian world and language are organized by contraries, that is, by a relation of *exclusion*: what is a bird cannot be a mammal, what is mortal cannot be immortal. Carving is dividing, forever; the distinct outcomes of the division are never to be reconciled. The impassable barrier between them is going to be signaled by the most characteristic word in this entire theory of the *logos*: the adverb "not." Negation is what gets analytic logic going. Let us get to the details of how this works.

The *logos* is meaningful discourse, we know; so, what is the most basic condition discourse must satisfy in order to be meaningful? Aristotle's answer is provided in

Metaphysics IV, in his defense of the Principle of Non-Contradiction (PNC). To begin with, he remarks that such a defense cannot take the form of a demonstration, as some principles are so fundamental that nothing more fundamental can be used as a premise in proving them:

> we have now posited that it is impossible for anything at the same time to be and not to be, and by this means have shown that this is the most indisputable of all principles.—Some indeed demand that even this shall be demonstrated, but this they do through want of education, for not to know of what things one may demand demonstration, and of what one may not, argues simply want of education. For it is impossible that there should be demonstration of absolutely everything; there would be an infinite regress, so that there would still be no demonstration. But if there are things of which one should not demand demonstration, these persons cannot say what principle they regard as more indemonstrable than the present one. (1006a)

(The reference to the undesirability of an infinite regress should be noted; we will go back to this topic in the next chapter.) And yet, though such people are not to be admitted into polite company, he will address their challenge anyway, if only by an indirect route:

> We can, however, demonstrate negatively even that this view is impossible, if our opponent will only say something; and if he says nothing, it is absurd to attempt to reason with one who will not reason about anything.... The starting-point ... is ... that our opponent ... say something which is significant both for himself and for another; for this is necessary, if he really is to say anything.... First then this at least is obviously true, that the word "be" or "not be" has a definite meaning, so that not everything will be so and not so.... And it makes no difference even if one were to say a word has several meanings, if only they are limited in number; for to each formula there might be assigned a different word.... [N]ot to have one meaning is to have no meaning, and if words have no meaning reasoning with other people, and indeed with oneself has been annihilated; for it is impossible to think of anything if we do not think of one thing; but if this *is* possible, one name must be assigned to this thing. Let it be assumed, then, ... that the name has a meaning and has one meaning; it is impossible, then, that being a man should mean precisely not being a man, if "man" is not only predicable of one subject but also has one meaning. (1006a–b)

Aristotle's negative demonstration goes on for much longer, but its main point should be clear by now: if we do not make radical distinctions, between predicates that cannot both be true of the same thing (at the same time), we indeed make no sense. What this not making sense *means*, on the other hand, comes dangerously close to a begging of the question: we do not make sense because everything ends up being the same ("the same thing will be a trireme, a wall, and a man, if it is equally possible to affirm and to deny anything of anything," 1007b), *that is*, because there are no longer any radical distinctions to be made; and as a result we are incapable of reasoning, *that is*, of practicing inference in the way Aristotle has taught us to do. But this feature of Aristotle's performance is only a reiteration (and evidence) of the fundamental character of PNC, which leaves no alternative for anyone who would defend it other than presupposing it, and reinforces the main point *I* am making here. In analytic logic, the exclusion asserted by PNC is the most basic necessary condition for the significance of discourse: if that exclusion fails, the integrity of this logic falls apart, and we no longer know what we are saying.

Before I proceed, a qualification is in order. Though the above is usually regarded as a negative demonstration of PNC, and I myself went along with this reading in presenting it, it is more directly relevant to contraries than to contradictories; for what it "demonstrates" we need to hold on to is our ability, once we apply one predicate, to reject the application of its contraries ("man," "wall," and "trireme" are contraries, not contradictories: a man cannot be a wall or a trireme, but something can be none of the above). I believe that this distinction is important and that the extent to which PNC and the attending notion of contradiction have monopolized attention is a mistake; but the monopoly is there, and I need to speak in such a way that I can be seen to clearly intersect common concerns. So, having had my say on the issue, for the remainder of this chapter I will accept the reduction of the rich notion of a contrary to the narrower one of a contradictory and speak as if PNC (not some variant of it, formulated in terms of contraries) were really the object of concern.

In addition to being the most basic condition of sense for analytic logic (any sentence, and specifically the conclusion of any argument, makes sense *because* it excludes its contradictory), PNC is also within this logic an invaluable reasoning tool. For any sentence that is shown to entail a contradiction must be denied—that is, itself excluded from the range of truths. How the tool is used has occasioned fiery debates within analytic logic. There seems to be no problem when the conclusion of this kind of argument (a *reductio ad absurdum*) is in fact purely negative. For example, no controversy arises from Euclid's proof that $\sqrt{2}$ is not a fraction. Euclid assumes that it is and also takes the fraction to be reduced to lowest terms; hence its numerator p and its denominator q must be either both odd or one odd and the other even. But then it is impossible that the square of p/q, that is, p^2/q^2, be 2. For, if p and q are both odd, then so are p^2 and q^2, and their ratio cannot be even. Similarly if p is odd and q is even. If p is even and q is odd, then, for some r, $p = 2r$ and $p^2 = 4r^2$, and there is no way that this, divided by an odd number, could have 2 as a result. The proof ends here, having refuted the original assumption, and we are given no clue on what $\sqrt{2}$ might be: the most we can say (and possibly more than Euclid himself would have wanted to say) is that it is one of the infinitely many things (more precisely: the infinitely many numbers) that are not fractions (I will return to this point in the next chapter).

But sometimes things are trickier. For example, early in his career, L. E. J. Brouwer proved that in every transformation (a function from a set into itself) a point remains fixed (goes into itself), by showing that a contradiction would follow from the hypothesis that there be no such fixed point.[1] Then, however, he became worried that the proof gave no idea of how to find this fixed point: no procedure for constructing it. And this worry was one of the driving forces in his development of intuitionism and in his later proposals for reformulations and new proofs of the fixed-point theorem.

---

[1] The complete formulation of the theorem includes conditions we can leave aside here, as they do not impact my current concerns: the transformation must be continuous and the set must be compact and convex.

What allows one to go from a purely negative conclusion of a *reductio* argument (such as Euclid's) to a positive one (such as Brouwer's) is the Principle of Excluded Middle

(PEM) *A* or not-*A*.

For, assuming that we have proved (as Brouwer did) it not to be the case that there is no fixed point (thus ruling out not-*A*), PEM lets us conclude that there is one (*A* must then be the case; there is, according to PEM, no third, or middle, option). Intuitionism, therefore, rejects PEM, which may well be a good idea in mathematics, if we think that every proof of mathematical existence should be associated with a construction procedure. But, however desirable and convenient such a requirement might be, it looks like a *special* one: whenever we claim that there is something, it requires that we know how to find it. And there are all sorts of non-mathematical things out there that we have no way of finding: the people responsible for various crimes, for example, of whose unknown identities and whereabouts (perhaps forever: think of Jack the Ripper) we must say, alas, *ça ne l'empêche pas d'exister*. Other objections to PEM come from other special contexts, where also we might want to avoid commitment to a definite yes-or-no choice: non-existent or fictional objects (Did Sherlock Holmes die on a Monday or not?), future contingencies (Will there be a sea battle tomorrow or not?), or category mistakes (Is the Atlantic Ocean thoughtful or not?). But Aristotle, for one, was quite sensitive to special circumstances, so much so that his formulation of PNC in *Metaphysics* IV is carefully worded to take care of them: "the same attribute cannot at the same time belong and not belong to the same subject in the same respect; we must presuppose, in face of dialectical objections, any further qualifications which might be added" (1005b).[2] So a lot of work needs to be done to show that similar provisions would not resolve apparent exceptions to PEM and that intuitionistic logic can be defended as a universal alternative to the dominant brand of analytic logic. In the absence of such work, we are left with what is suggested by the most elementary framework of analytic logic itself: barring special circumstances, if *A* and not-*A* are each all that the other is not, then everything must be one or the other. Which of course does not mean that radical alternatives to the analytic treatment of negation are not otherwise possible; and it is to them that I turn now.

In Chap. 3 I pointed out that negation does not exclude in dialectical logic. The opposite (*the contradictory*—which, as we are about to see, creates a serious problem) is the case: this logic thrives on overcoming negation. What was once small, hence not large, grows and becomes large; what was once mortal becomes immortal. Contradiction, far from being the most cogent reason for denying identity ("*a* is small and *b* is not, or *a* is mortal and *b* is immortal; so *a* must be distinct

---

[2] Note that, later in the same passage, Aristotle confirms the view I expressed above (and throughout) that contraries are more basic than contradictories: "If it is impossible that contrary attributes should belong at the same time to the same subject (the usual qualifications must be presupposed in this proposition too), and if an opinion which contradicts another is contrary to it, obviously it is impossible for the same man at the same time to believe the same thing to be and not to be; for if a man were mistaken in this point he would have contrary opinions at the same time" (1005b).

from *b*"), seems to be the engine driving dialectical identity (what it is to be something, dialectically understood): one's identity develops, one's truth comes to the fore, by constantly contradicting what one was. The matter, however, needs to be carefully reviewed, because a large amount of fuss has always been made within Hegel scholarship about whether Hegel's contradictions are retained (hence he is what is called a dialetheist[3]: one who allows for the coexistence of contradictions) or eliminated. The simple answer is (as often with Hegel) that they are both: the German word Hegel uses for what is to be done to contradictions is *"aufheben"* (together with cognates like the noun *"Aufhebung"*), a verb with the arresting peculiarity of including in its meaning both contrary operations of *canceling* and *preserving*. Hegel is delighted with this peculiarity:

> *"Aufheben"* has a twofold meaning in the language: on the one hand it means to preserve, to maintain, and equally it also means to cause to cease, to put an end to. Even "to preserve" includes a negative element, namely, that something is removed from its immediacy and so from an existence which is open to external influences, in order to preserve it. Thus what is *aufgehoben* is at the same time preserved; it has only lost its immediacy but is not on that account annihilated. The two definitions of *"aufheben"* which we have given can be quoted as two dictionary *meanings* of this word. But it is certainly remarkable to find that a language has come to use one and the same word for two opposite meanings. It is a delight to speculative thought to find in the language words which have in themselves a speculative meaning; the German language has a number of such. (1990, p. 107; translation modified)

Speculative thought is the one informed by dialectical logic, the one that recognizes identity in difference and in contradiction; so Hegel's delight in seeing his intuitions confirmed by ordinary language is comprehensible. Still, many will not share this delight and will think of *"aufheben"* as expressing an unfortunate ambiguity; so we must elaborate the issue further. (And, before we do that, I pause to notice that it has been a nightmare to find appropriate translations of *"aufheben"* in virtually any other, less "speculative," language. The common English translation "to sublate" (which I left out of the translation above) is a technical term made up for this purpose and unrelated (in the sense it would have here) to any non-Hegelian context; other candidates like "to overcome" and "to supersede" incline dangerously toward the extreme of canceling; I believe that the best translation—though one that, mysteriously, is hardly ever used by anyone else—is "to transcend.")

Return to the proof given in Chap. 3 of the non-Greekness of Plato. When the truth of this non-Greekness emerges in the narrative that is the meaning of "Plato," is his Greekness thereby dissolved? Is it no longer there? No: he is still the Greek Plato, born in Athens; and, whatever meaning we assign to his non-Greekness, it will have to coexist with his Greekness (the contradiction is retained). But, just as Plato has evolved within the narrative that defines him, so has every feature he once had, including his Greekness. What it is for him to be Greek has now gone through the same process of developing identity in difference; hence it has matured (say) from belonging to one of many highly contentious, divisive, and belligerent city

---

[3] This term was introduced by Graham Priest and Richard Routley in the first chapter of Priest et al. (1990).

states (what being Greek amounted to back then) to belonging to a site that is the cradle of rationality, the origin of the philosophical outlook, the overcoming (indeed!) of all ethnic invidiousness in the name of an enlightened understanding of humanity. And, if the narrative is cleverly constructed, it will be clear that it had to be those very contentious and divisive city states that could evolve into this paragon of enlightenment: that, say, it had to be the pressure constantly and painfully exercised by the Greek city states on one another (as opposed to the placid servitude prevailing in the Persian empire) that could eventually issue in the acknowledgment of the vanity of all such pressure and in the rational attitude that followed that acknowledgment. So it is precisely *as Greek* that Plato could transcend his Greekness, and his Greekness is never to be forgotten, though, in the new form it has acquired, it no longer contradicts his non-Greekness (the contradiction is also canceled). As Hegel would put it, his Greekness has added concreteness (or, as we might put it, conceptual detail, articulation) to his non-Greekness: what his non-Greekness now amounts to is substantiated, to a considerable extent, by his Greekness and by the *Aufhebung* of the contradiction between the two.

Along the same lines, if you were telling the story of a youth growing up (of what it means to be that youth), you might want to show that the hate he felt for his father, which at one point contradicted the love he felt for him (and made him prey to a perpetual, unresolved oscillation between the two feelings—which oscillation, to put it again in Hegel's terms, showed their identity in an immediate, still unintelligible way[4]), comes to be present in a more mature, and no longer contradictory, form of that very love, which has now incorporated an appreciation of the difficulties, frustrations, and faults common to the two of them and has grown into a hate of those faults and frustrations rather than of the man (though also still of the man insofar as he is taken to be an archetype of the faults and frustrations). In general, it is crucial in Hegelian logic that contradictions emerge in full force: that they threaten to break a structure apart and challenge it with explosion and destruction of its integrity—that they not be contradictions only in a manner of speaking, or by courtesy. And it is equally crucial that they be superseded in a later stage of reassurance, in which both horns of the contradiction, though they have changed, are still there (they manifest identity in difference) but (because of what change they went through) are no longer contradictory.

---

[4] In the same way in which, at the beginning of (1990), the oscillation of being and nothing present within becoming expresses in an immediate way the identity between the two, which Hegel (as I pointed out in Chap. 4) has just established: "*Pure being* and *pure nothing* are, therefore, the same. What is the truth is neither being nor nothing, but that being—does not pass over but has passed over—into nothing, and nothing into being. But it is equally true that they are not undistinguished from each other, that, on the contrary, they are not the same, that they are absolutely distinct, and yet that they are unseparated and inseparable and that each immediately *vanishes in its opposite*. Their truth is, therefore, this movement of the immediate vanishing of the one in the other: *becoming*, a movement in which both are distinguished, but by a difference which has equally immediately resolved itself" (pp. 82–83). A similar point can be made, for example, about the dialectic between the ground and the grounded to be found later in the same work (p. 447).

So negation is treated differently in dialectical logic and has a different impact on the items to which it is applied. Is it also the case that it is *itself* different than in analytic logic—that the word "negation" is ambiguous across the two doctrines? Some of Hegel's own terminology—especially when he talks about "determinate negation"—seems to support such a view. But I do not believe that the view is justified: I believe that the difference between analytic and dialectical logic, as I accounted for it, can explain the different roles that negation—one and the same negation, understood as an exclusion operator—has in them.[5]

Take an early phase of Plato's evolution: him as a twentysomething who follows Socrates around town. He is Greek, no question about it, and his Greekness denies his non-Greekness in the most obvious way. If the meaning of "Plato" were to be conceived analytically, as a collection of traits, then Greekness would be in and non-Greekness would be out, *and that would be the end of it*: Plato would be *determined*, among other things, by the negation of his non-Greekness (*omnis determinatio est negatio*). We would have problems understanding the continuity of Plato's identity across time, as with all spacetime continuants; but we would have such problems just because that is what his identity would consist of, within the analytic theory of meaningful discourse—which is the point I want to focus on now.

For this is the point dialectical logic rejects: in it, the meaning of "Plato" is a narrative that must proceed after the early phase we took up, and it must do so by transcending some of the traits that defined Plato in that early phase—by identifying in that phase the seeds of his development into a new phase in which he is defined by (among other things) the negation of those traits. We might say that, in following Socrates around town, Plato was already exposed to, and sympathizing with, a notion of humanity that denied ethnic idiosyncrasies, and that his later self-conscious acceptance of that notion was but the growing of the seed into a plant, the *truth* of it. Plato, then, denies his previous denial of non-Greekness (negation of the negation), and the second denial determines him positively as a new kind of human (determinate negation); but that is not because negation is different here—it is because determination is: because dialectical logic determines things, and words, differently, because it provides different meanings for the latter and different identities for the former. I said that, in dialectical logic, negation does not exclude; but that is not to be understood as saying that dialectical logic contains some "nonexclusive" negation. It is rather to be understood as saying that dialectical logic has a way of transcending the exclusion that negation brings in: of countering and neutralizing negation's divisiveness.

This discussion of negation instantiates a general point that deserves to be made here. Philosophers' theses are often radical ones: they propose radical changes in

---

[5] Of course, if the whole theory of what constitutes a meaning is radically different between different logics (indeed, it is constitutive of them and of their difference), we cannot expect negation to have the same meaning in analytic and dialectical (or oceanic) logics. But that will not be because of some specific difference concerning negation; rather, singling out some such specific difference will be a main tactic for *avoiding* recognition of the larger difference at stake among these theories. Just as, in the examples to be made below, introducing a verbal difference is a main tactic for defusing radical substantive differences among authors.

how we view things. A considerable amount of that radical character can be, and regularly is, defused by understanding their *words* differently—a practice that is greatly assisted by the translation of those words into some other language. Sometimes the only purpose served by such exercises is that of conveying an impression of higher academic dignity to the reader, as when the perfectly ordinary Freudian word "*Besetzung*," whose obvious English translation would be the equally ordinary "investment," receives the formidable rendering of "cathexis"—which, like "to sublate," is a technical term expressly coined for this purpose. But often there are, unfortunately, more serious philosophical diversions occurring. Frege's "*Bedeutung*," for example, is the most common German word for "meaning"; hence, when Frege says that the meaning of a sentence is its truth-value, he is saying something both substantive and controversial. When "*Bedeutung*" is translated as "reference," on the other hand,[6] that claim loses most of its edge: it gets reduced to a bland semi-definitional statement of whatever Frege might mean (or refer to?) by "reference" (and the operation borders on the ridiculous when the same word is translated—or, rather, transliterated—back into German and people start talking, in that language, about Frege's "*Referenz*"). Similarly, translators have been deeply embarrassed by Hegel's "*Begriff*": despite Hegel's own insistence that his *Begriff* was the same (dialectically, of course) as everyone else's,[7] they have typically refused translating it simply as "concept," preferring a capitalized "Concept" or a capitalized "Notion," which of course lays waste to the relevance Hegel should have to traditional conceptual concerns. The idea that Hegel might be utilizing his own special negation falls within the same general strategy, once again depriving Hegel's view (that negation must, and will, be transcended) of most of its force. And that is an analytic strategy of dividing and conquering: if pursued to an extreme, it would result in no two people ever disagreeing on anything—because they are just using words differently.

In turning to how negation works within oceanic logic, I find it useful to begin with Descartes' Third Meditation. There Descartes tries to prove God's existence, and his attempt is a clear failure, for various reasons—first and foremost, because "the natural light" can hardly deliver at this stage of the game the powerful premises the proof relies on.[8] But suppose we put aside this obvious case of the Cartesian

---

[6] With a certain amount of uncertainty and wavering. Max Black's translation of "Über Sinn und Bedeutung" flipped back and forth in its translation of "*Bedeutung*" in the various editions it went through, going from the original "reference" in Frege (1952) to "meaning" in Frege (1984), back to "reference" in Frege (1991; 3 years after Black's death).

[7] See, for example (1991a), p. 237: "because the concept has a meaning in the speculative logic that is so different from the one that we usually associate with this term, we might raise just the following question: 'Why is something that is so completely different nevertheless called concept?' For the result is that an occasion for misunderstanding and confusion is created. The answer to this question must be that, however great the distance between the concept of formal logic and the speculative concept may be, a more careful consideration will still show that the deeper significance of the concept is in no way so alien to general linguistic usage as it might seem to be at first sight" (translation modified).

[8] See, for example, this passage, where Descartes gives a much more conventional (and much less controversial) version of the principle he will ultimately use (see footnote 10 below), but still too

circle, we give Descartes the benefit of the doubt, and we follow the course the proof takes. It is a kind of cosmological proof, with a peculiar twist. The starting point is the idea of God; but, instead of proceeding as usual, by inquiring on the cause of that idea, and the cause of that cause..., Descartes short-circuits the whole process and gets to his conclusion in a single step.[9] For he distinguishes formal or actual reality (that is, reality, period), which pertains to everything that is, from objective reality (reality as represented), which only pertains to ideas, and then asserts the following principle: the cause of an idea must have at least as much formal reality as the idea has objective reality. Otherwise put, the cause of an idea must have at least as much reality, period, as the reality the idea represents.[10] A single application of this principle then gives him the desired conclusion: there are in his universe three degrees of formal reality—in ascending order, that of a mode, of a finite substance, and of an infinite substance. The idea of God, which has the actual reality of a mode, has, however, the objective reality of an infinite substance; therefore, by the principle he has asserted, its only possible cause is an infinite substance. It is, by all means, a valid argument; but why would we want to accept that principle? Surely there may be ideas whose causes have less than the exalted reality the ideas represent! We can think whatever we want—can't we?—but our thinking does not make it so, and does not necessarily come from something that is so!

Well, let us go slowly about it. First off, could the idea of a finite substance simply be caused by other ideas, and hence by modes? After all, we have many ideas of finite substances that do not exist—winged horses, chimeras, and the like. Couldn't they be generated without ever leaving the realm of ideas, with no input from substances, finite or otherwise? In fact, Descartes has already anticipated this objection, in the First Meditation, and here is what he says:

> it must surely be admitted that the visions which come in sleep are like paintings, which must have been fashioned in the likeness of things that are real, and hence that at least these general kinds of things—eyes, head, hands and the body as a whole—are things which are not imaginary but are real and exist. For even when painters try to create sirens and satyrs with the most extraordinary bodies, they cannot give them natures which are new in all respects; they simply jumble up the limbs of different animals. Or if perhaps they manage to think up something so new that nothing remotely similar has ever been seen before—

---

strong a version for him to be able to assert at the present stage of the meditating process: "Now it is manifest by the natural light that there must be at least as much <reality> in the efficient and total cause as in the effect of that cause. For where, I ask, could the effect get its reality from, if not from the cause? And how could the cause give it to the effect unless it possessed it? It follows from this both that something cannot arise from nothing, and also that what is more perfect—that is, contains in itself more reality—cannot arise from what is less perfect" (p. 28). It is hard to believe otherwise, to be sure; but it is also impossible to prove such believing immune, yet, to the threat of an evil genius.

[9] In the next chapter, we will see a more traditional variant of this proof—indeed, the original one, due to the Philosopher himself.

[10] "[I]n order for a given idea to contain such and such objective reality, it must surely derive it from some cause which contains at least as much formal reality as there is objective reality in the idea" (pp. 28–29).

> something which is therefore completely fictitious and unreal—at least the colors used in the composition must be real. (p. 13)

The idea of a chimera did not come from a chimera, as there is none, but it did come from a combination of the ideas of a lion, a goat, and a snake, which in turn came from lions, goats, and snakes. And even the idea of something stranger than a chimera, and not resembling anything at all, would still come from the ideas of something of that color, something of that size, something of that shape—all of which things (and lions and goats and snakes), though no chimeras, have the same formal reality as chimeras would have (they are all finite substances) and as the idea of a chimera has objective reality.

A similar response could be offered if someone objected that not just the idea of a chimera but a chimera itself—a sculpture of a chimera, say, or a genetically engineered one—issued from the idea of a chimera. Maybe so, but the idea of a chimera would still have issued from finite substances like lions, goats, and snakes—not to mention the fact that to go from the idea of a chimera to a bronze statue of one, or to a lab reproduction of one, would require the causal efficacy of finite substances like a sculptor or a geneticist. So far, then, Descartes' principle, however doubtful it might have looked at first, proves surprisingly resilient.

There is, however, another objection, which applies specifically to the ideas of infinite substances. Not all ideas, it seems, are either copied from some substance that was experienced or obtained by combining such copies. Some appear to be obtained by *negating* other ideas we have (and did acquire by experiencing substances). We experience finite rectilinear segments; we form the ideas of them; and then, by denying their finitude, we generate the idea of an infinite straight line, which we never experience and which cannot be obtained by combining ideas of finite rectilinear segments. We experience things (humans, the wind, fire) that have a certain limited power; and, by denying such limits, we generate the idea of something whose power is limitless. We experience things that last a finite time, and we generate the idea of something whose existence has no beginning or end. Isn't the idea of God obtained that way? And isn't its cause, therefore, a number of *finite* substances: the various finite things we experienced, and we ourselves as the deniers of their finitude?

Descartes has a response for this objection, too; but one that is far less convincing than his response to the previous one:

> And I must not think that, just as my conceptions of rest and darkness are arrived at by negating movement and light, so my perception of the infinite is arrived at not by means of a true idea but merely by negating the finite. On the contrary, I clearly understand that there is more reality in an infinite substance than in a finite one, and hence that my perception of the infinite, that is God, is in some way prior to my perception of the finite, that is myself. For how could I understand that I doubted or desired—that is, lacked something—and that I was not wholly perfect, unless there were in me some idea of a more perfect being which enabled me to recognize my own defects by comparison? (p. 31)

The step from the ontological priority of the infinite over the finite to its epistemic priority is a non sequitur (even granting that the finite is metaphysically grounded on the infinite, why should my thought of the finite be grounded on a thought of the infinite? and, even granting *that*, how does it follow then that my thought of the finite is grounded—not on a thought of the infinite, but *on the infinite*?); and, as for the one from my idea of lacking perfection to the existence of the idea of something perfect in my mind, it might have much greater plausibility, but it would not help his case—the point is not whether I have such an idea (I most certainly do) but whether a perfect being is the cause of it. Thus Descartes' principle fails precisely where he would want to apply it: not in relation to chimeras, or in general to finite substances, but in relation to God. And there is no way of rescuing Descartes from this failure, because he insists (as Spinoza did) that his argument is "geometrical" in nature, hence abiding by the same logic of Euclid: the logic Euclid got from Aristotle.[11] But perhaps there is a way of understanding what Descartes was implicitly and unconsciously driving at: a different way of thinking and reasoning altogether, quite similar to the one Spinoza (despite his own declared allegiance to the geometrical model) was going to instantiate later.

To see what might be going on, I bring up Bergson once again. In (1911c), he says:

> The very root of all the difficulties and errors with which we are confronted is to be found in the power ascribed here to negation. We represent negation as exactly symmetrical with affirmation. We imagine that negation, like affirmation, is self-sufficient. So that negation, like affirmation, would have the power of creating ideas, with this sole difference that they would be negative ideas.... We fail to see that while affirmation is a complete act of the mind, which can succeed in building up an idea, negation is but the half of an intellectual act, of which the other half is understood, or rather put off to an indefinite future. We fail to see that while affirmation is a purely intellectual act, there enters into negation an element which is not intellectual, and that it is precisely to the intrusion of this foreign element that negation owes its specific character. (pp. 302–303)

> *Negation ... differs from affirmation properly so called in that it is an affirmation of the second degree: it affirms something of an affirmation which itself affirms something of an object....* When we deny, we give a lesson to others, or it may be to ourselves. We take to task an interlocutor, real or possible, whom we find mistaken and whom we put on his guard. He was affirming something: we tell him he ought to affirm something else (though without specifying the affirmation which must be substituted). There is no longer then, simply, a person and an object; there is, in face of the object, a person speaking to a person, opposing him and aiding him at the same time; there is a beginning of society. (p. 304)

> whenever I add a "not" to an affirmation, whenever I deny, I perform two very definite acts: (1) I interest myself in what one of my fellow-men affirms, or in what he was going to say, or in what might have been said by another *Me*, whom I anticipate; (2) I announce that some

---

[11] In the *Second Replies*, pp. 113ff, he in fact provides a "geometrical" reformulation of this and other arguments, where the principle discussed above reappears as Axiom V: "[T]he objective reality of our ideas needs a cause that contains this reality not merely objectively but formally or eminently" (p. 116). And Axiom V provides the crucial step in the proof of Proposition II (the other steps being fairly obvious), which is the thesis proved in the Third Meditation: "The objective reality of any of our ideas requires a cause which contains the very same reality not merely objectively but formally or eminently (Axiom V). But we have an idea of God (Def. II and VIII), and the objective reality of this idea is not contained in us either formally or eminently (Axiom VI); moreover it cannot be contained in any other being except God himself (Def. VIII). Therefore this idea of God, which is in us, must have God as its cause; and hence God exists (Axiom III)" (p. 118).

other affirmation, whose content I do not specify, will have to be substituted for the one I find before me. Now, in neither of these two acts is there anything but affirmation. The *sui generis* character of negation is due to superimposing the first of these acts upon the second. It is in vain, then, that we attribute to negation the power of creating ideas *sui generis*, symmetrical with those that affirmation creates, and directed in a contrary sense. No idea will come forth from negation, for it has no other content than that of the affirmative judgment which it judges. (pp. 305–306)

The claim articulated in these passages would do the trick for Descartes in responding to the objection above: if no new ideas are generated by the use of negation, then that is not the way the idea of God could have come to us. It would only be of interest for Bergson scholarship if this claim were just something Bergson believed—a "principle" he adopted. What makes it more interesting than that, and even more relevant to Descartes, is that we have already identified Bergson as (implicitly) espousing oceanic logic. So the question arises: is it possible that this view of negation be a consequence *for both of them* of the adoption not of some specific "principle" but of oceanic logic itself? Could it be that, because they think and reason according to this logic, both find it natural to view negation as they do?

What might make it difficult to answer these questions is the fact that oceanic logic was introduced here by way of sorites, and sorites are typically formulated by utilizing negation. No definite number of grains of sand makes a heap, so there are no heaps. The picture conjured up by this formulation of the argument is that there are two predicates, each negating the other—to be a heap and to be a non-heap—and we have no way of determining how far one or the other applies. The same seems to be the case for *bald* and *non-bald*, *white* and *nonwhite*; or, for that matter (if we include Zeno's paradoxes), for *moving* and *non-moving*, *reachable by Achilles* and *non-reachable by him*, *getting to the target (by the arrow)* and *not getting there*, etc. So, one will ask, how is negation not playing a major role here in generating new ideas? How can the above be reduced to affirmations only?

These formulations are not, however, what oceanic logic is about. They represent an instrumental use of intuitions that are at home within oceanic logic in the building of weapons that are meant to undercut its main rival—analytic logic. These weapons are supposed to be *paradoxes* (let us not forget that); they are *incredible*, *absurd* conclusions, *beyond belief*; but they have nothing incredible or absurd for someone reasoning oceanically. They do for someone reasoning analytically. It is the latter who will think that anything must be either a heap or not a heap, either bald or not bald; reasoning by exclusion belongs to him, and it is in order to make trouble for him that oceanic logic brings up negation and shows its rival to be incapable of using it consistently. In the same fashion, a dialectical logician might point out that Socrates must be both young (at 20) and not young (at 70), to show that analytic logic cannot reason comfortably about spacetime continuants; though, when it comes to developing her own way of reasoning, there is a lot more she will say than this simple (and, in her view, simpleminded) opposition.

If we put aside the use of sorites as polemical tools and think of them as suggestive of another theory of the *logos*, then our best starting point will be one I mentioned before: the spectrum of colors. You start with *orange*, say, and at some (indefinite) point you find it more natural to refer to what you are looking at as *red*. You could, of course, also think of what happens then as negotiating the passage from *orange* to *non-orange*; but then it would be obvious that you are introducing an external perspective, one that is not sensitive to the reality of the situation and drastically limits any understanding of it. The reality of the situation is that anything in that portion of the spectrum could be seen as either dark orange or light red; and, if you go further, it could be seen as *very* dark orange or as just plain red; and how you see it will depend on your point of view; and the same will continue to be true as you go toward green or blue; but none of that richness could be captured if you just put it in terms of what is or is not the case.

This is how we should think of heaps, too, if we think oceanically. We should get away from dividing up the world into heaps and non-heaps; make a spectrum, instead, that includes globs and lumps and heaps and piles and hills and mountains; and realize that any aggregate of grains of sands could be any of the above, depending on the perspective brought to it (what is a glob for a human could be a mountain for an ant), and that the reason why a title like *The Englishman Who Went Up a Hill and Came Down a Mountain* is amusing (as is the movie by that title[12]) is that any attempt at determining precisely what is one or the other is going to be felt as an arbitrary imposition, a mindless pigeonholing of those nuances that make the world a constant object of discovery, and constantly worthwhile.

Before I proceed, this is as good a time as any to face what some will inevitably find it hard to accept. The inevitability comes with the territory: the way we think and reason shapes our opinions more deeply and irresistibly than any data could; it provides the framework within which all data will be received and understood; so, for all those who see no way of thinking and reasoning other than analytically, a lot of what I am saying here will make no sense. Still, attempts must be made at helping them come around: examples must be offered that might make them click and see a duck rather than a rabbit—shift, that is (which can only be an instantaneous event, however long and strenuous the process was that prepared for it), from their ordinary understanding of whatever is given to a different one.

Getting to the point, it is going to be hard for most people, when faced by a blue and a green patch, to accept and take seriously the question, "Which is the most orange?"; but then you might try two different tacks. On the one hand, you might ask them to regard this exchange as part of a parlor game and, if they agree to play, you might point out that, in the course of parlor games, people continue to think and reason, and that a theory of the *logos* should be able to capture how they think and reason then. (By doing this, you will try to involve them in the play of logics I discuss in Chap. 11.) On the other hand (and returning to a suggestion made in Chap. 4), you might also show them two white faces and ask them, "Which is the most black?", or two black faces and ask them "Which is the most white?"; and nowadays they might feel that they have to answer (this time because contemporary politics has put *them* in play); then you could ask them how these questions are different

---

[12] From 1995, directed by Christopher Monger and starring Hugh Grant. In the movie, the arbitrary imposition and the mindless pigeonholing are carried out by cartographers.

from the previous one (after all, black-and-white movies can now be colorized). Neither strategy is guaranteed to work; but guarantees are the wrong thing to look for, when one is trying to bring about a conceptual revolution.

Returning to the main topic of discussion, Bergson's firm rejection of negation as generating new ideas (and his invocation of difference as a substitute for negation) and Descartes' more confused (and confusing) arguments to the effect that the idea of an infinite substance could not arise by the negation of finitude make sense in the context of a commitment to oceanic logic (which of course neither of them, as indeed no one else so far, expressed clearly, but which we can see at least implicitly made by Bergson). Analytic logic is defined by exclusion, understood as an ineluctable verdict. In dialectical logic, exclusion is a perpetual stimulant, inviting us to go to work in overcoming it (to scratch where it itches) and reach a stage in which we show ourselves (to ourselves as much as to others) inextricably implicated in what we once excluded, at one with it. In oceanic logic, exclusion has no currency. We are always talking about the same thing, incapable of describing its true essence—of doing full justice to it—but still voicing how that essence strikes each of us, how each of us sees it from where she is sitting. Fixating on a single view would be the most unproductive attitude: as the totality of views accessible to us is still limited, restricting that totality (possibly to one element only) would make our limitations even more serious and even more of an impediment. This kind of fixation is what we get when we exclude: when, instead of adding another, different view to the general conversation, we hold one view *at the expense of another* or *as incorporating another*, or, even worse, we simply try to explode one of the views.

In Chap. 4, when introducing oceanic logic, I asked you to picture a wave "looking powerful, running fast and tall toward you," and to "identify for a moment with it, imagine it to be self-conscious and proud of its power and speed, to feel indestructible and immortal." However exhilarating that sentiment might be, it originates in exclusion: in thinking of oneself *as opposed to* what is not oneself—as identical to that wave but not to other ones, nor to anything else. And it comes to an abrupt end when the wave crashes and is reduced to foam. To think of the ocean (and of the ocean of being, as Spinoza would) as only variously wavy is to think of any wave we might be as just another pretty festoon that decorates the ocean's surface, without fathoming its depths. And it is to think that the silliest thing we might do is to announce the perfect faithfulness of what is projected by a particular wave to the reality of those depths, the absolute transparency of the depths within the texture of that wave.

A few years before Descartes and Spinoza, Tommaso Campanella reversed millennia of epistemic predominance of the sense of sight (as the one, I noted before, which senses from a distance and does not modify the object sensed) by asserting the priority of taste (as the sense that most integrates the subject with the object sensed), and said that we know the world as a worm knows cheese.[13] That, too, is a

---

[13] "Sapience ... comes from sapor, and from the taste that feels things as they are. This sense gives sapience its name more than the others, since it perceives not only the external and strong natures of things, but also the intrinsic, hidden, and weak ones. Touch knows in wine the cold that from the

position germane to oceanic logic. The worm experiences the same mysterious world as the eagle does, and the best both could do in order to know better would be to compare notes; but the eagle is likely to think that its experience is more perspicuous than the worm's, *to be preferred over the worm's*; whereas all it is doing then is being blind to its perspective *as a predator*.

None of the above is going to save Descartes' argument in the Third Meditation. For that argument is offered as an inference, hence it is supposed to shove its conclusion down his readers' throats. As such, it is a miserable failure. But the above does suggest that, lurking in the back of Descartes' mind, there might have been another attitude altogether, which inclined him to take a dim view of negation. There is more to the object of thought than any thinking, more to the origin and substance of the world than any experience of it; so let us not quickly draw barriers between what can and what cannot be, based on the limitations of our viewpoints. Let us conceive of infinity as the inexhaustible, *affirmative* source of countless specifications and of our own thinking as reflective of this inexhaustible process, to an extent: of the ideas that come to our minds as reflective of some portion of it and of the idea we have of the source itself as a sign that, if only as a worm lodged in cheese, we resonate with it. (Also, let us think of that very source—of the object of theology, if that is what it is to be—as something whose *difference* from any of our descriptions of it will always escape us, *not* as something that will invariably *deny* those descriptions: as something immeasurably richer than those descriptions, not exclusive of them.) That the *Meditations* end with a reference to how consistency, at the limit, would turn dreams into wakefulness might be an additional clue that oceanic logic is indeed in the background of Descartes' reasoning (dreams are not just opposed to wakefulness; they slowly graduate into it by manifesting more and more consistency), as was to be in the forefront of Spinoza's; but we are not engaging in historical scholarship here. What matters for us is that negation—understood as we always did, as an exclusion operator—plays no role in oceanic logic. Infinity does, in a way in which it does not in either of its competitors; so it would seem that there is where we should direct our attention next.

---

outside takes hold of it and detains its forces; taste, instead, feels both the external cold and, through the sapor, the internal hidden heat, which is the intrinsic heat that arises when what is ingested is crumbled, or it flows penetrating the tongue's pores, hitting the spirit of its nerves, and manifesting itself to them.... Only this sense, therefore, becomes intrinsic to the object, and the whole object to the sense; hence the one is sapient who knows things from the inside and not superficially by the intellect" (1994, p. 75; my translation). "[O]ur acquaintance with man, wine, and our things is sufficient for a generic knowledge of them, which is necessary for the preservation of life, as for the worm it is sufficient to be acquainted with the small parts of cheese in which it was born and on which it feeds, but it is not sufficient to know simply what are wine, bread, man, and their origin, and their elements, and the world and the one who governs it. In the same way the worm does not know the whole cheese, or its essence, or its place in the world, or the elements and causes from which it comes" (ibid., p. 67; my translation). I often ponder what would have happened to Western civilization if the views of Campanella (who died in Paris in 1639, highly revered by Louis XIII and after having predicted—he was also an astrologer—the birth of the Sun King), and specifically his low opinion of the intellect and his preference for a knowledge that was sensual and embodied (suggested in the first passage above), had prevailed over those of Descartes (whose two major works, the *Discourse* and the *Meditations* were published in 1637 and 1641, respectively). Or maybe I should say, in light of the discussion in the text: if Campanella's views had prevailed over Descartes' self-conscious, official ones.

Chapter 8
Infinity

Abstract For Aristotle, actual infinity is an absurdity: it is absurd to think of something that has no end and also comes to an end. (The cosmological argument, which he initiated, is based on this view.) Conventional lore has it that Cantor's set theory, developed within the framework of post-Aristotelian analytic logic, offers a systematic, positive account of infinity—or, rather, of infinities, because there are infinitely many of them. But careful examination of the key Cantor's Theorem shows that all that set theory can assert about infinities are *infinite sentences*, which have an affirmative form but are empty of positive content.

In dialectical logic, infinity is not a question of size but consists in the overcoming of limits. Every moment in which one recognizes one's identity with one's other is a moment of infinity; conversely, even a large infinite set that was set in opposition with one of its subsets would count as finite insofar as it was limited by that opposition.

In oceanic logic, infinity is given; indeed it is the only thing that is given. It is not by chance that the best examples of oceanic logic in the history of philosophy come from reflection on infinite, real entities: God for Anselm and Spinoza, life for Bergson, being for Heidegger. And it is affirmative infinity, because (as explained in Chap. 7) here negation has no currency. Leaving extraordinary contexts and concerns aside, Zeno's paradoxes prove us to be implicated in infinity during our most mundane transactions, and prove the vanity of all our efforts to colonize infinity, to put labels on it and organize it in workable ways.

Aristotle regards a *reductio ad infinitum* as equivalent to a *reductio ad absurdum*. Though there is in his corpus nothing comparable, concerning infinity, to his defense of PNC,[1] he clearly abides by the principle that, if a position entails that an infinity be *actual*—that is, given, completed, existent as such—then that position must be

---

[1] As I pointed out at the time, his rejection of infinite regress shows up in the course of his "arguing" for PNC. The rejection invoked there, however, is quite reasonable, as it concerns the very structure of analytic arguing: if an analytic argument is intended to provide a justifying ground for a claim, the ground should not constantly give way—it should come to a resting point—or we will never in fact have justified the claim. It is harder to understand why there could not be an ontological infinite regress (which is what I discuss in this chapter): why the series of antecedent causes of an event (as opposed to the series of antecedent premises of an argument) should not be infinite.

rejected, just as if it entailed a contradiction. For the notion of an actual infinity is incoherent: infinity can only refer to a process that never ends, like that of adding units to a number or of dividing a segment into parts; and it is incoherent to think of something that both never ends and also comes to an end and delivers a completed outcome. That, by the way, is how he came upon the cosmological argument, whose Cartesian variant I discussed in the previous chapter: if anything is or happens, then the whole series of its causes must be actual or the thing would not have been or happened, and, since nothing actual can be infinite, there must be a first member of it—an uncaused cause of it all (or, alternatively, a cause of it all that is also *causa sui*). Here is a relevant passage from the *Physics* (the issue is also taken up in the *Metaphysics*):

> If then everything that is in motion must be moved by something, and by something either moved by something else or not, and in the former case there must be some first mover that is not itself moved by anything else, while in the case of the first mover being of this kind there is no need of another (for it is impossible that there should be an infinite series of movers, each of which is itself moved by something else, since in an infinite series there is no first term)—if then everything that is in motion is moved by something, and the first mover is moved but not by anything else, it must be moved by itself. (256a)

It might be useful to remark that there can be no corresponding argument for the conclusion that the unfolding of causes and effects must come to an end, since that unfolding can take place *potentially*: one step at a time, *making actual* one effect at a time, but never being *entirely* actual. Whereas there cannot be a merely potential cause of something actual: if something is actual, all of its causes must be actual as well.

The very Greek word for infinity, *ápeiron*, is indicative of trouble. The *ápeiron* is the unbounded, the formless, the chaos from which the world, according to traditional mythology, originated. No rational thinking is possible in the *ápeiron* and of the *ápeiron*; indeed, what the world's origin amounted to was the introduction of an orderly, *finite* structure—the *cosmos*—which the rational mind can penetrate, understand, and cognize. Allegiance to this structure was hard to let go, and challenges to it were not well received: Giordano Bruno was imprisoned, tortured, and burned at the stake for claiming (among other things) that the universe is infinite.[2] In mathematics, Carl Friedrich Gauss still regarded infinity, at the beginning of the nineteenth century, as a *façon de parler*, since the only sense he could make of it was the potential infinity allowed by Aristotle (the infinity of a process that never ends) and any assertion of a potential infinity is, in fact, a disguised assertion of finitude. The cash value of a sentence like

---

[2] "[T]he Nolan ... holds that the universe is infinite, whence it follows that no body can simply be in the middle of the universe or at its periphery or anywhere between these two limits except through certain relations to other nearby bodies and artificially imposed limits" (1995, p. 152). And he holds that his view is more respectful than the (then) orthodoxy of the true nature of God: "not only ... this philosophy contains the truth, but also ... it supports religion better than all other kinds of philosophies, such as those which suppose the world to be finite, the effect and efficacy of divine power limited, the intelligences and intellective natures only eight or ten, the substances of things corruptible, the soul mortal" (ibid., p. 182).

8 Infinity

(1) There are (potentially) infinitely many prime numbers

is nothing other than

(2) For every *finite* prime number, there is a larger *finite* one

Conventional lore has it that this adamant opposition was reversed (at least as far as mathematics was concerned) beginning in the 1870s, first with various proposals for the definition of an irrational number—all entailing references to actual infinities[3]—and then, most impressively, by Georg Cantor's set-theoretical treatment of transfinite numbers, defined by David Hilbert "the most admirable flower of the mathematical intellect and in general one of the highest achievements of purely rational human activity" (1967, p. 373). But this edifying story deserves a closer look.[4]

To begin with, note that, from a logical standpoint, the most obvious expression of the Aristotelian attitude toward infinity is *infinite sentences*: sentences phrased as affirmatives but containing a negative predicate, an example of which is

(3) The rose is non-white.

(In the tradition, which includes Kant,[5] these were known as infinite *judgments*, since the emphasis was on the act of judging them by the mind. I will continue to

---

[3] In Dedekind's definition, for example, an irrational number is defined as a *cut* in the field of the rational numbers, that is, an ordered pair of two actually infinite sets—understood intuitively as that of all rational numbers smaller than, and that of all rational numbers larger than, the irrational number to be defined.

[4] Cantor's project proceeded concurrently with the one pursued by Frege that I discussed briefly in Chap. 6, and Frege's "logic" (as I suggested then) was inclusive of the same naïve set theory Cantor was working on (though it made use of it for the different goal of proving the logical nature of arithmetic). So, when paradoxes hit, they threatened to bankrupt both projects. Hilbert, as we have seen, rose in defense of set theory, but only in the form in which it was developed by Cantor; as is clear from his correspondence, he had no high opinion of Frege.

[5] It is instructive to quote here part of Kant's discussion of infinite judgments in (1998), as it suggests points made later in my text: "in a transcendental logic *infinite judgments* must also be distinguished from *affirmative* ones, even though in general logic they are rightly included with the latter and do not constitute a special member of the classification. General logic abstracts from all content of the predicate (even if it is negative), and considers only whether it is attributed to the subject or opposed to it. Transcendental logic, however, also considers the value or content of the logical affirmation made in a judgment by means of a merely negative predicate, and what sort of gain this yields for the whole of cognition. If I had said of the soul that it is not mortal, then I would at least have avoided an error by means of a negative judgment. Now by means of the proposition 'The soul is non-mortal' I have certainly made an actual affirmation as far as logical form is concerned, for I have placed the soul within the unlimited domain of undying beings. Now since that which is mortal contains only one part of the whole domain of possible beings, but that which is undying the other, nothing is said by my proposition but that the soul is one of the infinite multitude of things that remain if I take away everything that is mortal. But the infinite sphere of the possible is thereby limited only to the extent that that which is mortal is separated from it, and the soul is placed in the remaining space of its domain. But even with this exception this space still remains infinite, and more parts could be taken away from it without the concept of the soul growing in the least and being affirmatively determined" (A72–73 B97–98).

use the term "sentence" for them since I treat them as I did all their counterparts before and will hereafter—as purely linguistic entities.) "Infinite" here conveys (consistently with Aristotle's view) the message of something indefinite, imprecise, uninformative: though we appear to say what the rose is, we are not; we are throwing the rose into the infinite, chaotic cauldron of those things that—whatever they might be—are not white. Which is confusing; truly affirmative sentences like

(4) The rose is white

say something distinct about the rose, and negative sentences like

(5) The rose is not white

exclude the rose, just as distinctly, from the range of white things, without saying anything more because nothing more is to be said. Infinite sentences, on the other hand, trick you into thinking that something distinct is being affirmed of the rose (because of the form the sentence takes) but leave you with no new knowledge of it.

Now consider the following statement of set theory:

(6) There are uncountable sets.

(6) is what turns the issue of actual infinity from a subject of philosophical preference (or prejudice) into a respectable, even a venerable, topic. If all the infinity we had were the denumerable one of natural numbers, then it would appear entirely optional whether to think of these numbers as given one at a time or all at once (infinity could continue to be, as was for Gauss, a manner of speaking); but, if we can prove that there are *different* infinities, some larger than others, then a whole new area of inquiry has suddenly opened up—the paradise from which Hilbert claimed no one shall be able to drive us. So we can understand how, of all the theorems of set theory Cantor proved, the one that allows us to assert (6) has come to be known as *Cantor's theorem*.

Let us get into this proof, then. For every set A, define $\mathbf{P}(A)$, the *power-set* of A, as the set of all subsets of A. We also define a set A *larger* than a set B ($A > B$) if there is a one-one correspondence between a subset of A and B (a function that associates each member of that subset of A with exactly one member of B), but there is no one-one correspondence between B and (all of) A. There is a one-one correspondence between every A and a subset of $\mathbf{P}(A)$: the subset constituted by all the unit sets $\{a\}$ of members a of A—just make every a correspond with its unit set. Is there a one-one correspondence between any A and (all of) $\mathbf{P}(A)$? Suppose, by *reductio*, that there is; call it f. Then, for every member a of A, it is either the case that $a \in f(a)$ or not. Define the set $X = \{x: x \notin f(x)\}$. X is a subset of A; so, for some member a of A, $X = f(a)$. Question: $a \in X$? Suppose that it does; then a belongs to the set of all members x of A such that $x \notin f(x)$; so it doesn't; therefore, $a \notin X$ follows from its own contradictory and is proven true. Suppose, on the other hand, that it doesn't; then a does *not* belong to the set of all members x of A such that $x \notin f(x)$; so it does; therefore, $a \in X$ follows from its own contradictory and is proven true. We have thus proved the contradiction $a \in X \;\&\; a \notin X$ and the hypothesis that there be an f establishing a one-one correspondence between any set A and $\mathbf{P}(A)$ has been refuted.

Therefore, for every set A, $\mathbf{P}(A) > A$. This is Cantor's theorem. In particular, the power set $\mathbf{P}(N)$ of the set N of natural numbers is larger than N: it is not finite and it is not denumerable as N is—it cannot be exhausted by counting all of its members one by one, either for a finite time or forever. It is uncountable. Furthermore, the same theorem shows that the power set $\mathbf{P}(\mathbf{P}(N))$ of $\mathbf{P}(N)$ is larger than $\mathbf{P}(N)$, and that *its* power set is larger than it, and so on forever—and that all of them are uncountable.

When we assert this conclusion, there is something we intend to say: some sets are so large that the ordinary operation of counting, even if extended to infinity, will not, indeed, *account* for them; they will always contain residues that are not reached—and those residues, to make the claim even more sublime and terrifying, are as infinitely large as the whole sets themselves. The real numbers are uncountable,[6] but so are the real numbers that are irrational (that, like $\sqrt{2}$, are not ratios of natural numbers), and so are the ones that are transcendental (those that, unlike $\sqrt{2}$ but like π or like the base of logarithms e, are not roots of algebraic equations); so, to put it simply (and most impressively), we cannot count the numbers that we have no systematic way of naming—the unknown is bottomless. The flip side to this sublimity and terror, however, is that set theory seems to give us a way of *handling* the bottomless unknown with all the precision of a mathematical theory. We can assign numbers (however uncountable) to transcendence and can run ordinary operations on them. We can compute the results of adding them and subtracting and multiplying and dividing them and elevating them to whatever power. We can order them in an infinite hierarchy; we can say what the successor of one of them is; we can distinguish between those that have a predecessor and those that do not—that can only be reached as the limit of a series. It truly seems as if we have conquered paradise, as if we have colonized the *ápeiron*, and, without denying its infinity, we can say all sorts of definite things about it. Or, rather, about *them*, because there are an infinity of infinities.

But is what we say really definite? Gödel's theorem tells us that we could never prove the consistency of set theory (unless we accept an even stronger theory): we could never, therefore, prove that the theory makes sense—that it will not prove to be its own *reductio* and collapse, like Frege's, by the discovery of a contradiction. We have to take it on faith. Suppose we do. If the theory is consistent, then (by the crucial result that Leon Henkin 1949 proved, on his way to establishing the completeness of first-order logic) it has a model; and if it has a model, we are informed by the Löwenheim-Skolem theorem;[7] it has a denumerable model. Whence the Skolem paradox: how can a theory like that of sets, which asserts that there are uncountable, hence non-denumerable, sets, have a denumerable model? How can a model as large as N include sets, like $\mathbf{P}(N)$ (not to mention $\mathbf{P}(\mathbf{P}\ldots)N)$), that have been shown to be larger than N?

---

[6] For it can be proved that there is a one-one correspondence between $\mathbf{P}(N)$ and the set R of real numbers. Also, to make sense of what follows, both rational and algebraic numbers are denumerable, so what is left by subtracting them from the reals is not.

[7] See Löwenheim (1967) and Skolem (1967a).

To answer these questions, we must go back to the proof of Cantor's theorem. In order to show that **P**(A) is always larger than A, the proof relies on the definition of "larger than," as applied to sets. The definition has two parts. Forget about the first one: that there must be a one-one correspondence between a subset of **P**(A) and the whole of A. Concentrate on the second: there is *no* one-one correspondence between the two (whole) sets. Piece of cake, then: all we have to do is make sure that no such correspondence exists, and, no matter how *intuitively* large the sets involved are, one will be, in the model, larger than the other. If the second one is denumerable, the first one will not be: it will be uncountable.

A model of a theory T is a structure that verifies T: that makes all of its axioms true and all of its rules truth-preserving, hence all of its theorems true. Because model theory is defined within set theory, a structure verifying T is a set D (the *domain*) on which an *interpretation i* is defined of all the nonlogical elements of (the language of) T: a function that assigns to all its constants members of D; to all its predicates subsets of D (if the predicate is monadic) or sets of n-tuples of members of D, for $n > 1$ (if the predicate is n-adic, for $n > 1$); and to all its n-ary function symbols functions from D^n into D. Now suppose that D is denumerable. Intuitively, there is not enough stuff in there to verify Cantor's theorem; for the largest any set can be in D is denumerable, so how can a model based on D have anything representing, say, the set R of real numbers, which we know to be uncountable? Well, all it takes is to so define i that no function exists establishing a one-one correspondence between N and R (while one exists that establishes a one-one correspondence between N and a subset of R)—however N and R are represented. But, one might protest, *we know* that in a denumerable domain such a function could always be defined; so how could it not exist? The answer is as firm as it may be frustrating: a model is a mathematical entity, and intuitive claims to knowledge have no currency in determining it—only its (actual, not just possible) definition does. If set theory is consistent, then it must have a model; if it has a model, then it must have a denumerable model; therefore, in that model words like "uncountable" will not have the meaning we ordinarily take them to have—they will just mean what the definition of "uncountable" requires: that a certain function does not exist. And if, in the mathematical entity which a model is, the relevant function does not exist, everything is as it should be.

The Skolem paradox has been milked for all sorts of philosophical morals.[8] The one I want to extract from it is as follows: within the analytic logic that constitutes the framework of set theory, sentences presuming to say something definite about infinities do not; all that they are is infinite sentences—hence there is nothing definite about them. To reiterate, a sentence like

(7) *R* is uncountable

intends to say something of vast significance about R: that R has a power, a cardinality, vastly superior to that of N and that the real numbers are vastly more than the naturals. And that may well be the case in set theory's *intended model*: the

---

[8] Most famously by Hilary Putnam, starting with his (1980)

structure most mathematicians have in mind when they go to work in set theory. But intentions do not make reality: what is meant is not always what is said (a point Hegel would have appreciated[9]). If set theory is consistent then it has a model, we know; and, if it has a model, it has uncountably many non-isomorphic ones, including uncountably many where the domain is denumerable and hence "uncountable" cannot possibly mean what we *intend*. So a more honest, transparent way of expressing what (7) says is

(8) R is non-countable.

That is, R is one of the indefinitely many things that we cannot count, in all the indefinitely many circumstances in which we might find ourselves and whatever we might understand as "counting" there.

One more remark is in order, before leaving the topic. Though infinite sentences exist relative to finite domains, they have the most poignancy, and do the most damage, when (as in the case of set theory) the domain is infinite. For suppose I say:

(9) I am non-Chinese.

(9) says nothing definite about my nationality, but it can be easily turned into something that does. For there are only 200 or so nationalities that one can be, so saying (9) is in fact equivalent to saying:

(10) I am Italian or French or Spanish or …,

where, importantly, the ellipsis indicated by the three dots *can* be filled in, by listing all the (finitely many) nationalities that are not Chinese. So we need to take another look at (3): the reason why that sentence is "throwing the rose into the infinite, chaotic cauldron of those things that—whatever they might be—are not white" is that there are (uncountably?) infinitely many members of the spectrum of colors, so a disjunction corresponding to (10), in this case, could never be completed. If instead there were, let us suppose, only one million colors, the affirmative form of that infinite sentence *could* be taken seriously: the sentence would in fact be affirming something distinct (however tenuously disjunctive) about the rose. In that case, the finiteness of the domain would do the job of turning the in(de)finite sentence into a definite one; but within an infinite domain this kind of rewriting is impossible, and hence the relevant sentences are going to stay in(de)finite.

In conclusion, analytic logic is incapable of saying something affirmative about infinity. By a judicious use of *reductio* arguments, it can prove sentences that appear to make such affirmations: there are uncountable sets; there are uncountable levels of uncountability; there is a transfinite hierarchy of infinite cardinals (all uncountable except the first one). And we may be inspired by such claims and believe that infinity has been conquered, until we realize *what* they are actually claiming: how everything they say could be true in the old-fashioned, undistinguished, and (for all we know) *potentially* infinite realm of natural numbers. The intended model of set theory may be all that is needed to do mathematics effectively; as far as logic is

---

[9] Remember the quote in footnote 13 in Chap. 3.

concerned, we need to make clear what those intentions are worth and how there is nothing more to them than a vague gesturing at the *ápeiron*. Instead of the alleged contact with an infinity of infinities, what we have here is a whole lot of *sentences* about infinity, a lot of *talk* of infinity; and, however consistent that talk might be, it does not, in and by itself, make infinity definite.

Moving now to dialectical logic, begin with some quotes from Hegel:

> the genuine infinite ... consists ... in remaining at home with itself in its other. (1991a, p. 149)

> the peculiar quality of spirit is ... to be the true infinite, that is, the infinite which does not one-sidedly stand over against the finite but contains the finite within itself as a moment. (1971, p. 23; translation modified)

> The will which has being in and for itself is *truly infinite*, because its object is itself, and therefore not something which it sees as *other* or as a *limitation*; on the contrary, it has merely returned into itself in its object. (1991b, p. 53)

Infinity, here, is not a question of size. Finitude is limitation; and, whenever limitation is denied (negation of the negation), infinity is asserted (determinate negation). Plato might have thought that his Greekness stood opposed to all non-Greeks (*hoi bárbaroi*); and, however large the set of Greeks might have been, even if there had been more Greeks than stars in the sky or drops of water in the ocean, that opposition made it finite. Infinity is attained when Plato realizes that the other (the non-Greek) is the same as himself, when he acquires familiarity with the other, when he feels at home there. Not because he is giving up on himself, but because he is acquiring an ampler view of what "himself" is. If there is a sacrifice going on, it is not a sacrifice of Plato, but of his previous isolation: the blood that is being shed is the one that always accompanies a birth—in this case a *re*birth, a turning of Plato's previous caterpillar into a beautiful butterfly, no longer stuck on a surface but ready to float in the air and connect with all sorts of things under the sun. Even in the extreme case of infinite sets, and leaving aside all the complications we explored above, if an infinite set stands opposed to one of its subsets, maybe an infinite one, neither set is truly infinite: either is limited by the other, insofar as it denies it and takes itself to be what the other is not; hence both sets are finite. To manifest infinity, both sets would have to be regarded (regard themselves?) as (dialectically) identical to one another.[10]

For analytic logic, infinity is what surrounds the *cosmos*: it may be godly or demonic, but in all cases it exceeds our understanding. We venture into it at our own risk: the words we utter about it might suddenly prove meaningless, as the semantics of the divine reveals itself incommensurable to our own or the mathematics of it slips into incoherence. (So it is true of analytic logic, too, as was said above of oceanic logic, that its only possible theology is a negative—or differential—one. In this sense, they both oppose dialectical logic.) Not surprisingly, then, Hilbert tried

---

[10] "If the finite is limited by the infinite and stands on one side, then the infinite itself is also something limited; it has its limit in the finite; it is what the finite is not, it has something over against it, and it has its limit and boundary in that" (Hegel 1984/1988 I, p. 293).

to reconcile us with this challenge by rephrasing it in terms of metamathematical finitude—and couldn't. For dialectical logic, on the other hand, there are moments of infinity everywhere within the *cosmos*, within the orderly arrangement of our experience; and they are the moments when we make progress in our understanding and transcend opposition. I may have thought that my own person contradicted my neighbor's: that we had conflicting interests, and the only possible relation between us was a struggle. When I overcome that view of myself and my neighbor, when I realize that I can only successfully pursue my interests by pursuing them together with him, by turning them into *our* joint interests, when my *I* gives way to a *we* in which I am reconciled with him, then I have an experience of infinity and make *this* world, the small world of my neighborhood, which could only be regarded as finite in comparison with galaxies or real numbers, less scattered, more integrated, more in touch with itself, *more infinite*. It will continue to be finite to the extent that it is still opposed to another neighborhood; but now I know what I am supposed to do with that residual finitude—with any finitude. I know that it is not a matter of adding more items, as when I add one more natural, or prime, or real number and find myself contemplating the same sort of organization I had before: I know that this potential infinity is also *spurious* infinity, that it reproduces itself, and its limitations, forever and ever.[11] I know that I have to face the opposition, work things out with it, and do the same with all other oppositions I will run into; I know that this is what infinity is all about.

Giacomo Leopardi, the greatest Italian lyrical poet, wrote *L'infinito*, one of his masterpieces, in 1819, at the age of 21 (and 3 years after the publication of Hegel's *Science of Logic*). A literal English translation of it reads:

Always dear to me was this solitary hill
and this hedge, which from so many parts
of the far horizon the sight excludes.
But sitting and gazing, endless
spaces beyond it, and inhuman
silences, and the deepest quiet
I imagine in my thoughts; where almost
my heart sinks. And as the wind
I hear rustling through these trees, I that
infinite silence to this voice
keep comparing; and I feel the eternal,
and the dead seasons, and the present
and living one, and the sound of if. So in this
immensity drowns my own thought
and foundering in this sea is sweet to me.

---

[11] "Something becomes an other, but the other is itself a something, so it likewise becomes an other, and so on ad infinitum. This *infinity* is *spurious or negative* infinity, since it is nothing but the negation of the finite, but the finite arises again in the same way" (Hegel 1991a, p. 149).

The reason why this poem is a masterpiece of philosophical reflection as well as of metric and melodious enchantment (where the latter is better appreciated in the original, of course) is that two distinct visions of infinity convene in it. Infinity is the immeasurably large, the unfathomable, the scary *ápeiron*, that had by then been the subject of inquiry and debate for at least a century under the guise of the *sublime* (which I evoked above and which had received an extended and insightful treatment in Kant's third *Critique*[12]): there belong the endless spaces, the inhuman silences, the deepest quiet so much present to the mind of an author (Leopardi) one of whose first works, at the age of 15, had been a history of astronomy. But note how infinity enters the scene: it is a modest hedge, on a hill that is close to home and is a frequent destination of his walks, that strikes him as *excluding* most of the view that would otherwise be accessible from there. The hedge is a limit to his perception and, as Hegel would say, to recognize something as a limit is already to think of what is beyond it, on the other side of the limit: it is already an *Aufhebung* of the limit, an announcement of the limitless. For an analytic logician, the sight of the hedge would simply be an extrinsic occasion for bringing up the true nature of infinity: of the infinite universe that astronomy (and Leopardi) delves into and that, once properly focused on, will make us let go of the hedge that suggested its existence and relevance (the hedge would only belong to the heuristics of infinity). For a dialectical logician, all the talk of immensity and eternity is but code for what infinity is all about, for how spirit becomes aware of it (and of itself as infinite): through seeing an ordinary object as a barrier, as a border, and hence accessing what that very ordinary object, suddenly, shows up as a step into—the overcoming of any barrier. The poem straddles this divergence and ambiguously refers to both parties of it, which makes for its unstinting fascination as an object of irrepressible intellectual and emotional play—an artistic jewel. For us, here, it is both a terse reminder of how differently analytic and dialectical logic conceive of infinity and an anticipation of the play among theories of the *logos* to be discussed in Chap. 11.

In oceanic logic, infinity is not the dumb, forlorn other, and it is not to be painstakingly conquered by repeated acts of supersession. It is given—more than that: it is the only thing that is given, and recurs invariably after each attempted encasing of it within a neat definition. After all, the examples I used to introduce this form of thinking center on entities that are both infinite and absolutely real: God for Anselm, God again (or substance or nature) for Spinoza, life for Bergson, being for Heidegger. Analytically, infinity is elusive: we have signs for it—alephs, omegas, and the like—but they hardly express more than our *desire* for infinity. As we skate on the thin ice of using contradiction to prove features of it, we never know that contradiction is not already infesting our proofs; and, even if it didn't, the best we can hope for is negative lessons—lessons in how *not* to think of infinity (uncountable, inac-

---

[12] I will return to Kant's sublime in the next chapter. For the moment, let me notice that he gives identical characterizations of the sublime and of the infinite: "*That is sublime in comparison with which everything else is small*" (2000, p. 134). "The infinite … is absolutely (not merely comparatively) great. Compared with this, everything else (of the same kind of magnitude) is small" (ibid., p. 138).

cessible, unnameable, etc.). Dialectically, infinity is posited in the affirmative; but its actuality is always that of making this or that *Aufhebung*—since spirit is endlessly creative, we cannot hope to get to the end of it. The impression of a completed process is an aftereffect of the logic's retrospective stance: wherever one is in the dialectic, one is at the end of the process so far, and her task is one of rationalizing the process so far, of making it sound as inexorable as it is providential. The end of history is only ever the end of the path that spirit has traveled until now. Oceanically, however, infinity is the first thing we have and the one we return to in the wake of every attempt at characterizing it, and of finding the characterization to fall short. Infinity is affirmative here not because negation is denied but because (as was explained in Chap. 7) negation has no currency; so the one, infinite reality whose every image is distorted must constantly affirm itself and constantly prove itself slashed and curtailed by all those (finite) images.

If the infinity of *everything*, or of the foundation of everything, is the most congenial habitat for oceanic logic, we must not forget that this is a general doctrine of meaningful discourse; so infinity must show up in all contexts, wherever we think and understand things that way. And it does. Zeno's paradoxes prove us implicated in infinity during our most mundane transactions: as we get from the couch to the door, as we pass another car on the freeway, as we observe a bird's flight. And they prove the vanity of all our efforts to colonize this infinity: well-meant efforts, well thought out, even useful ones; and yet, evidence of an empty laboring for attaching labels that will not stick, for trying to have something settle in a reassuring mold that has nothing reassuring about it, and will continue to explode all molds. Infinity constantly proves itself far less than reassuring, and yet as the most familiar environment: *unheimlich*, and yet tasting like home.

Early on, I talked of oceanic logic as tending to agreement. Now, in light of what followed that early appraisal, we must elicit the fact that agreement is itself a vexed notion. In analytic logic, agreement is reached by working from the same premises to the same conclusions and excluding any statement that is contrary to those conclusions. In dialectical logic, it is reached by transcending the occurrence of contrary views: by coming to see them as different stages in the evolution of the same view. In oceanic logic, it is reached by admitting, to oneself and others, that the matter of contention escapes us all, that we are all indeed just making honest efforts at capturing what will not let itself be captured, what will appear to each of us foreshortened by his own private slant; so we might as well integrate our slants because only a more complex view, one that coalesces all that is seen from the most disparate angles, will have us go in the right direction—will get us closer to a richness, and a complexity, that cannot itself be scaled to manageable size. In oceanic logic, infinity is there at the beginning and always wins at the end; what is negative, delusive, to be rejected is the finite: not because it is the particular item it is, for that is as good as any other, but precisely because it is finite. And we must aim at agreement, and at compromise, among ourselves because anything each and every one of us will ever be able to figure, and utter, is a finite, hence inevitably unfaithful, projection of the infinite that necessarily eludes us.

Chapter 9
Mathematics

Abstract Mathematics is the most natural terrain for analytic logic; indeed, one way to think of analytic logic is that it conceives everything mathematically. Still, the other logics have an important, though generally unrecognized, presence within this discipline.

The best way to see mathematics dialectically is to start from a consideration of its history: of its gradual expansion of the concept of number from the rationals to the irrationals, then to imaginary and complex numbers, then to transfinite ones. An analytic logician might appreciate all of this but would insist that the history of mathematics is external to mathematics: if *there are* irrational, complex, and transfinite numbers, then those numbers exist eternally; their existence has no history, though there is a history to humans coming to know them—a history of mathematicians, then, not even quite a history of mathematics. This view, though perfectly consistent, is however itself *internal* to the analytic point of view: what is in question here is the identity of numbers, and whether this identity is construed analytically or dialectically is an issue that cannot be decided independently from a commitment to one or the other logic. For dialectical logic, the history of mathematics (understood logically, not chronologically) simply *is* mathematics.

The role of oceanic logic within mathematics can be best illustrated by focusing on two issues. First, recursive definitions: characterizations of what something (say, a natural number) is that indefinitely postpone the moment of fulfillment and hence never conclusively establish the distinction between what is to be defined and its contraries. Set theory allows us to prove a recursion theorem establishing that the clauses of a recursive definition identify exactly one function, but for the practicing mathematician exactly the same problem will surface in set theory itself, since the theory's intended model (within which the practicing mathematician works) is itself defined recursively. The second issue is continuity, which analytic logic has a tendency to construe in terms of discrete entities, thus violating what from an oceanic standpoint is its very nature.

Mathematics is the terrain where analytic logic finds itself most naturally at ease. Mathematical objects are, with no exception, "what they are and not another thing"; mathematical proofs establish definitive results, which drastically rule out any other possibility. While ultimately critical of mathematical knowledge (in that its definitiveness does not go quite far enough), Plato recommended a full immersion in mathematics for those who would eventually see the forms[1]—analytically perfect entities plagued by no ambiguity or vagueness. An unsympathetic, but not inappropriate, understanding of this recommendation is that he was proposing to brainwash the future rulers of the state, so that they would come to see all everyday experience as if it were mathematics—though mathematics as we know it (I said) ended up for him being still unsatisfactory and another, more adequate science (a more thorough brainwashing: what he called dialectic) was needed. Contemporary mathematical scientists often seem to encourage the same understanding of what they do: especially mathematical economists, when they blame people for not being good enough (rational enough) in matching their mathematical models. One wonders what would happen if they were made kings; and one is often horrified when they become, at least, counselors to kings.

Given this state of affairs, the present chapter will be structured differently from previous ones. The three theories of the *logos* will not be taken up in turn and accorded (more or less) equal space. Rather, I will assume that reasoning mathematically is mostly reasoning analytically and try to show how, in spite of such analytic prominence, dialectical and oceanic logics are not absent from this area of concern.

The natural way of getting dialectical logic into the conversation is by introducing a narrative element into mathematics, which declares itself and its claims to be foreign to any development. And it is easiest to do so by addressing issues from the history of mathematics—though this straightforward term will be called in question below and reveal itself a major object of contention.

In Chap. 7 I rehearsed Euclid's proof that $\sqrt{2}$ is not a fraction. I could also say (and did say later) that it is not a rational number; let us think for a moment about this expression. What it means is that $\sqrt{2}$ is not the ratio of two natural numbers; but "rational" is a loaded word. We might *mean* only the above; but we are also *saying* that $\sqrt{2}$ is not something that reason can make sense of—that, as a number, it would be an absurdity. This is exactly how Euclid, and the Greeks in general, took the proof: not that $\sqrt{2}$ was not some kind of number, but could be some other kind; $\sqrt{2}$ was not a number at all; it could not be; it was unreasonable to think that there could be such a number and a number theory (an arithmetic) that dealt with similar monstrosities. Because the geometrical equivalent of $\sqrt{2}$ stared at you from geometrical diagrams (say, from the diagram of a square of side 1 and diagonal $\sqrt{2}$), geometry was judged to have vastly higher expressive powers than arithmetic and was regarded as the paradigm of rigorous thinking for the next 2000 years.

---

[1] Ten years of it, according to the *Republic* 537–39. For the relevance of this study, see, for example, the following: "we ought to … persuade the people who are going to undertake our community's most important tasks to take up arithmetic. They shouldn't engage in it like dilettantes, but should keep at it until they reach the point where they can see in their mind's eye what numbers really are, and they shouldn't study it as merchants and stallholders do, for commercial reasons, but for the sake of warfare and in order to facilitate the mind's turning away from becoming and towards truth and reality" (ibid. 525b–c).

Eventually, irrational numbers were allowed, and the conjunction of rational and irrational numbers formed the field of the real (another loaded word!) numbers; how did that happen? It happened as people slowly turned their attention from what numbers were ("the odd and the even," to speak Plato's language, and their ratios) to what could be done with them. We have seen this to be the case with transfinite numbers, and what we said then can be repeated now: you may want to call a series of digits divided by a decimal point, infinite after the decimal point, and possibly falling into no finite pattern after the decimal point a "number," but that will just be your arbitrary choice, or your opinion, until you prove that you can work with these "numbers" just as you do with the rational ones—that all arithmetical relations apply to them and all arithmetical operations can be carried out on them. Then, if you have successfully refocused everyone's attention away from the metaphysics of numbers and onto this pragmatic aspect of them, and persuaded everyone that this is what matters about numbers, you will be able to realize a monumental *Aufhebung*: a dialectical development in the very meaning of "number," which asserts its total continuity with what numbers always were while immensely expanding their range.

At some point, the metaphysics of numbers was going to become an object of interest again, and people like Dedekind and Frege (among others) were going to get busy trying to sort it out; but that was not before the emphasis on pragmatics had made the theory of numbers such metaphysics would have to account for grow beyond any expressiveness geometry might ever claim, and reverse centuries of priority of spatial intuition over abstract thought. For example, we prove that there exists no such thing as $\sqrt{-1}$, don't we? That is, we prove that there exists no number that, multiplied by itself, gives -1 as a result. Nothing prevents us, however, from turning this negative outcome into a positive proposal—into making $\sqrt{-1}$ a nonexistent, unreal, *imaginary* number—and then showing how to work effectively with those *complex* numbers which contain both a real and an imaginary portion. And, because a complex "number" is nothing other than an ordered pair of reals, why not generalize on this strategy and talk about *quaternions* as legitimate subjects of mathematical study, or *matrices*? Of course, transfinite numbers are also coming, and at this stage of the narrative they will be seen not as metaphysically suspect but as a clear stroke of genius.

Whoever reasons analytically will consider the above irrelevant to mathematics proper. Mathematics, he will insist, deals with entities that are out of time (and space, though some of them have spatiality); therefore, properly speaking, mathematics does not have a history. What exists mathematically and what is mathematically true is eternal. What does have a history is humans' belabored access to mathematical existence and truth. So, assuming that there exist, in fact, irrational and imaginary and complex numbers, and that what we take today to be true of them is in fact true of them, there is no history to any of that: it was always true and always will be. There is, however, a history of how humans came to appreciate such facts: to become aware of truths that had been there all along. The "history of mathematics," then, should receive another name, to avoid confusion: "history of mathematicians," say, or even more explicitly "history of human approaches to mathematics."

But, as often happens when logics confront one another, this claim of irrelevance is internal to the analytic point of view and is senseless when evaluated outside that context. To be sure, the analytic identity of number cannot be expected to change— that is why spacetime continuants are such a problem for this notion of identity. But the dialectical identity of number, and of mathematics itself, is a different story; indeed, *it is a story*, not an immutable formula, and conflicts and contradictions are to be part of it, as are their resolutions. Within this other context, what it is to be a number, or what it is to be mathematics, will be some such narrative as the one I told; and it will be crucial to the narrative providing meaning that something that could not possibly, at some stage, be a number, hence a subject-matter for mathematics—something for which it would have been absurd, at that stage, to think that it could be—transcend this impossibility and become the pivot for a new, revolutionary phase of the identity of number, and of mathematics. In dialectical logic, the history of number, or of mathematics (to be understood as a thoroughly connected plot, not as aimless chronicle), constitutes number, or mathematics, itself.

In Chap. 6, I described the analytic horror of paradoxes that pervaded mathematics (and the philosophy of it) in the first few decades of the XX century. It might be productive to take another look at that episode, from the perspective of the "dialectical mathematics" I have alluded to now. Then those paradoxes might look like the growing pains of a new dialectical development that would have mathematics mutate into—what? Dialectical logic cannot be used to make predictions: its stance, we know, is a retrospective one. And, as I pointed out in that earlier passage, the enthusiasm and the thrust that animated the reflection on mathematics back in the days have waned, and overweening, overarching programs and manifestoes are sorely missed; so it is impossible to fathom what, if any, mutation will occur. But I can at least express (as I did before) the wish that it be a mathematics, and a science, that has learned to live with the paradoxes: that has learned how all complex enough structures have paradoxical outcomes, yet we don't want to give up complexity in order to be safe. Democratic freedom must face the paradox of those who would want to use their freedom to impede the freedom of others; languages allowing self-reference must face the liar. But we would not want to give up democratic freedom or (*pace* Tarski) self-reference; rather, we will face the paradoxes when they arise and isolate them from our more ordinary dealings. We might also come to the point (I wish we did) where mathematics coexists with paradoxes, acknowledges them to be evidence of richness and complexity, and quarantines them when they come up, without for that reason forfeiting any of its substance and rigor.[2]

---

[2] In a conversation with Moritz Schlick on December 30, 1930 (as reported by Friedrich Waismann in *Philosophical Remarks*), Wittgenstein seems to endorse this attitude: "**WITTGENSTEIN**: If, when I was working with a calculus, I arrive at the formula $0 \neq 0$, do you think that as a result the calculus would lose all interest? **SCHLICK**: Yes, a mathematician would say such a thing was of no interest to him. **WITTGENSTEIN**: But *excuse* me! It would be enormously interesting that precisely this came out! In the calculus, we are always interested in the result. How strange! This comes out here—and that there! Who would have thought it? Then how interesting if it were a contradiction which came out! Indeed, even at this stage I predict a time when there will be mathematical investigations of calculi containing contradictions, and people will actually be proud of

9 Mathematics

To explore the role that oceanic logic has in mathematics, return to numbers once more: to the simplest and most primitive ones—natural numbers. In 1889 (at a time Frege was busy completing his *Grundgesetze der Arithmetik*) Giuseppe Peano axiomatized arithmetic.[3] His system contained nine axioms. Four stated basic properties of equality (hence were analogous to Euclid's common notions). The remaining five (corresponding to Euclid's specifically geometric postulates) read:

(1) Zero is a natural number.
(2) If n is a natural number, then $s(n)$ (the successor of n) is a natural number.
(3) For every two natural numbers m and n, if $s(m) = s(n)$ then $m = n$.
(4) For every natural number n, $s(n) \neq 0$.
(5) For every set K, if $0 \in K$ and if, for every natural number n such that $n \in K$, $s(n) \in K$, then, for every natural number n, $n \in K$.[4]

(3) and (4) are simple statements, much like those used by Hilbert in his 1899 axiomatization of geometry.[5] (1) and (2), however, are trickier. To begin with, they do not state something to be the case about natural numbers; they state, rather, what natural numbers are. More than axioms, they look like definitions. But—and here comes their second, and most relevant, oddity—they are not the sort of definition that we have become used to from analytic logic and that are current in other branches of mathematics, as when we state

(6) Two lines on a plane are parallel = $_{df}$ they have no point in common,

or even in arithmetic

(7) A prime natural number = $_{df}$ one whose only divisors are 1 and itself.

---

having emancipated themselves even from consistency" (Wittgenstein 1975, p. 332). Apparently, as opposed to dialectical logicians, Wittgenstein believes he has a crystal ball—one that confirms their general framework.

[3] In (1889), a book whose first few lines read auspiciously: "The questions that pertain to the foundations of mathematics, though they have been treated by many people these days, still lack satisfying solutions. This difficulty arises mostly from the ambiguity of discourse. For this reason it is of the greatest interest to devote attention to the very words that we use. I have embarked in this examination, and in the present work I present the results of my study, and their applications to arithmetic" (p. iii; my translation). The logic of arithmetical discourse, in other words, will tell us a lot about how to address arithmetical problems: a more hopeful task, as it turned out, and one more germane to my own here, than Frege's attempted reduction of arithmetic itself to logical discourse.

[4] I have made inessential changes to Peano's formulations to allow for easier comparisons with contemporary accounts. The most significant one is that for Peano 1, not 0, is the first natural number.

[5] But note (anticipating a point to be made later in my text) that, in the introduction to Hilbert (1903), he describes his task as follows: "The present inquiry is a new attempt to set up a complete and consistent simple system of axioms for geometry and to derive from them the most important geometrical theorems in such a way that the *meaning* [*Bedeutung*] of the different groups of axioms ... may clearly come to light" (p. 1; my translation and emphasis). So his axioms are understood by him as implicit definitions.

That is, they do not define a natural number by genus and differentia. What they do instead is build the set N of natural numbers by an endless process: N contains 0 (by (1)); and, because it contains 0, it also contains 1 (by (2)); and, because it contains 1, it also contains 2 (by (2)); and... Which means that this definition of N—called *recursive*—will never be completed: the dots will always be an essential part of it; you know that 0 is in there, and 1, and 2..., and after a while you get the idea. But is "getting the idea" the way to do rigorous mathematics? Frege, for one, with his utter disdain for the role of intuition, would not have approved.

One could get rid of (1) and (2) by accepting Hilbert's proposal that some terms be regarded as primitive and that the entire axiom system amount to an *implicit* definition of those primitives. I mentioned Hilbert's axiomatization of geometry above; in it, we find axioms like

(8) There are at least four points not lying on a plane

but Hilbert will not tell you what a point or a plane are—you can take a point or a plane to be *whatever* satisfies the axioms pertaining to those primitive terms (famously, he said that "[o]ne must be able to say at all times—instead of points, straight lines, and planes—tables, chairs, and beer mugs"[6]). Following this advice, we could say that there is no need to define natural numbers by (1) and (2): natural numbers are whatever satisfies *the remaining* Peano axioms.

However, it's not so easy. For take (5) now. It contains a reference to a set K, which means that, properly speaking, it does not belong in number theory but in *set* theory: a much more powerful affair (and a highly questionable one—think of the paradoxes that were to be discovered in it). We could try to hide this fact by rephrasing (5) in terms of properties, that is, by replacing it with:

(9) For every property P, if P holds of 0 and, for every n, if P holds of n then P holds of $s(n)$, then P holds of every n.

But a property of natural numbers, from the extensional standpoint that governs mathematics, is nothing other than a set of them; so the distinction between (5) and (9) is purely verbal. If we want to clearly separate out *elementary* arithmetic (arithmetic without an infusion of set theory), we need to go the way Skolem did and reformulate (5) as an axiom *schema*[7]: a collection of infinitely many axioms having the general form.

(10) If $A(0)$ and, for every x, $A(x)$ entails $A(s(x))$, then, for every x, $A(x)$.

---

[6] See Reid (1996), p. 57.
[7] See Skolem (1967b). I must add, however, that, though an axiom schema allows for an easier comparison with (5) and (9), Skolem himself uses the equivalent form of a *rule*: if A is proved true of 0 and, assuming it is true of x, it can be proved of $s(x)$, then A is true of every x. Also, he regarded (11)–(14) below as *definitions*—consistently with my forthcoming comments. Finally, note that, from the very title of his article, he refers to a recursive *mode of thought*.

9 Mathematics

And, to get started with some basic statements about natural numbers from which (by using (10)) to be able to derive all others, we need to characterize addition and multiplication:

(11) For every x, $x + 0 = x$
(12) For every x and y, $x + s(y) = s(x + y)$
(13) For every x, $x \times 0 = 0$
(14) For every x and y, $x \times s(y) = (x \times y) + x$.

Which brings us back to the same situation we thought we had managed to avoid. I cautiously said that (11)–(14) "characterize" addition and multiplication; but, speaking bluntly, they tell us (try to tell us) what addition and multiplication are—they define them. And we will never get to the end of these "definitions." By repeated applications of (11)–(12), we will be able to say what the sum of 14 and 27 is; but we will never be able to say *what a sum is*. And isn't that frustrating, if what you are axiomatizing (providing a systematic account of) is *arithmetic*?

Again, there will be those who are not impressed. Every sum we ever face is the sum of two specific numbers, and (11)–(12) tell us how to compute it. Which is fine, if individual sums are all we care about. If, on the other hand, we care about what it means to be a sum, as I do in this book,[8] then the following is food for reflection: It is not just that analytic logic does not allow for everything to be defined; Plato and Aristotle were already aware of that, and Hilbert's suggestion to think of an axiom system as an implicit definition of all its primitives is as good as any. It is, rather, that the axiomatization of arithmetic, no matter how we phrase it, asks us to "characterize," or define, certain notions in a peculiar way—as what is attained at the end of an endless process. The recursive definition of a natural number, or of addition and multiplication, works much like a sorites: by indefinitely postponing the moment of fulfillment, it effectively shows that definition is never attained.

If something like a sorites recurs within the most elementary mathematical practice, then oceanic logic has a place in understanding it: numbers, natural or otherwise, might be something that this practice variously tries to "characterize" without ever coming to the bottom of it. Far from being the most precise, explicit, and rigorous of human practices, mathematics might embody the very human striving, and the very human frustration, of dealing in earnest with a reality that forever escapes us—as it forever escapes us how blue is different from green or how an arrow can be moving at all. But, if this is how the practice works, it is not how the most common ideology of that practice, which is couched in analytic terms, sees it and describes it. When arithmetic is "reconstructed" within set theory, one proves a *recursion theorem*, establishing (once and for all!) that the clauses of a recursive definition such as (11)–(12) identify exactly one function; thus addition is proved to

---

[8] At the end of (1967b) Skolem justifies his "definitions" by saying that his work "is built upon Kronecker's principle that a mathematical definition [*Bestimmung*] is a genuine definition if and only if it leads to the goal by means of a *finite* number of trials" (p. 333). Maybe so, but the problem remains of what the goal is: defining every individual sum or defining the operation of sum?

exist and, in the language of set theory, it can be defined in the manner Aristotle demanded—by genus and differentia. We are home, finally. Or are we?

Reconstructing another mathematical theory within the foundational theory of sets amounts to justifying it: making it conform to the dominant ways of thinking and reasoning. But then, after you completed this ideological task, you will have to engage with set theory itself—to work in it. As a mathematician, you will be guided by what you take the universe of sets to be, by your intended model of set theory. I mentioned the intended model in Chap. 8, while making a different point; now let us rather simply ask ourselves, "How is this model defined (or 'characterized')?" The illuminating answer is that what we get here is a generalized form of a recursive definition, more intricate than that of natural numbers because it extends to the whole hierarchy of transfinite (uncountable—whatever *that* means) sets but, for our purposes, not substantially different—not different, to be specific, in how it indefinitely postpones the moment of fulfillment. Here, too, we start with something like 0—the empty set { }—and proceed endlessly with an analogue of the operation of successor: the power-set operation. *At the end* of this endless pursuit, we collect everything we have generated so far (well, we do not, really, since *we* never come to the end of it; but we suppose that *there is* an end to this endless process) in a first infinite set, ω; then we start again with the power-set operation; and, at the end of another endless pursuit, we again collect all we got; and so on. The construction looks like this:

{ }
{{ }}
{{ }, {{ }}}
{{ }, {{ }}}, {{ }, {{ }}}
.
ω
{ω}
{{ω}}
{{ω}, {{ω}}}
.
2ω
.
ω^2
.
ω^ω
.
.
.

An uncharitable (but, for us, relevant) conclusion to be drawn from this situation is that insistently, and irritatingly, mathematical practice, as opposed to the logic we superimpose on that practice, keeps on foregrounding another logic: one in which infinity is the elusive target of a failing effort, not a secure object of possession. Insistently, oceanic logic proves itself relevant to mathematics.

In light of this discussion, it is instructive to return to potential infinity. So far, we have seen it only negatively: as the negation that infinity exists (is actual)—infinity as only a *façon de parler*. The affirmative view, then, would be the one asserting that infinity is actual: a view that sees, as unlikely allies, Cantor and Hegel, who defended actual infinity two generations before the birth of set theory, in three extensive remarks of the *Science of Logic*.[9] Of course, actual infinity meant something different for the analytic and dialectical perspectives represented by those two authors; what begins to surface now is that potential infinity can also be understood differently, depending on the logic you use. Analytically, it is a manner of speaking; dialectically, it is spurious; either way, it is *avoiding* infinity, looking away from it. But, oceanically, it is *pointing to* infinity: to a reality that can only ever be suggested, intimated, announced, but never attained *and therefore can only be experienced through this intimation of it*.[10] Hegel's main polemical objective, in declaring potential infinity spurious, was Kant's philosophy, where actual infinity is downright contradictory but infinity is nonetheless inexorably addressed, thematized, sought for—where the sublime representation of what cannot be represented both reconciles us with our rational destiny and makes that destiny the goal of an interminable journey.[11] One way to look at the dispute between Kant and Hegel is as a dispute

---

[9] On pp. 240–313. (These remarks taken jointly, or even each of the first two taken separately, are by far the longest in the whole book.) See, for example, p. 248: "the infinite series contains the spurious infinity, because what the series is meant to express remains an *ought-to-be* and what it does express is burdened with a beyond which does not vanish and *differs* from what was meant to be expressed. It is infinite not because of the terms actually expressed but because they are incomplete, because the *other* which essentially belongs to them is beyond them; what is really present in the series, no matter how many terms there may be, is only something finite, in the proper meaning of that word, posited as finite, i.e., as something *which is not what it ought to be*. But on the other hand, what is called the *finite expression* or the *sum* of such a series lacks nothing; it contains that complete value which the series only seeks; the *beyond* is recalled from its flight; what it is and what it ought to be are not separate but the same."

[10] I said in Chap. 6 that negative theology provides a model for the attitude oceanic logic has to any of its objects; then, in Chap. 7, I requalified that negation as difference, and in Chap. 8 I contrasted the negative/differential attitude of both analytic and oceanic logic toward infinity with the positive one of dialectical logic. Here we see a point of contrast between the theologies germane to analytic and oceanic logics, and one that gives more content to the distinction between "negative" and "differential." For analytic logic, what is at issue is negation as sheer exclusion: if we say that God is not, for example, this tree, we are only excluding the tree from the purview of the meaning of "God"—and, one exclusion after the other, we end up excluding *everything* we are familiar with from that meaning. For oceanic logic the tree, though not bringing our quest for God to a close, might be a pointer for the direction that quest should take—much as the collapse of the distinction between black and white might be a pointer for the direction our quest for humanity should take or the collapse of the distinction between Category 4 and Category 5 storms might be a pointer for the direction our quest for a proper way of coping with natural disasters should take. The tree might be a name for God not in the sense (current in analytic logic) of a label attached to an object we have experience of but in the sense of a sign, an arrow that shows which way to go in order to approach (though we might never attain) that object. As was suggested in Chapter 7, the sign will always be irremediably different from the object, but not quite a negation of it.

[11] See Kant (2000), p. 134: "nothing that can be an object of the senses is … to be called sublime. But just because there is in our imagination a striving to advance to the infinite, while in our reason

between two logics—which, insofar as they could only understand anything in their own terms, could not understand each other. Kantian reason is as unrealizable (in both senses of this word) as God is for the believer; and yet anything (a river, a flower, a person dear to us) could be a name of God, a finger pointing in his direction. Same with reason: every step we make in our moral progress, every instant that we live as if there were to be a future for us, and, yes, every number that we add to the store of what we have already counted make infinity show up in our experience, not because any of that approximates it (there is no getting closer to it; it is always infinitely far) but because it makes experience into something that receives its meaning from infinity. Something that, in pure oceanic speech, can be seen (and see itself) as a *mode* of infinity.[12]

There is another area of mathematics where oceanic logic is relevant, not impertinently manifesting its presence but, just as impertinently, voicing its dissatisfaction. It has to do with the recurrent confrontation between arithmetic and geometry. One of the main elements of this confrontation is that arithmetic deals with *discrete* entities, neatly divided from one another, whereas geometry deals with *continuous* ones, which merge into one another.[13] Through a process that took most of the XIX century and came to be known as the *arithmetization of analysis*, mathematicians reconstructed the calculus (whose original formulation was deeply influenced by geometrical intuitions) in terms of numbers and eventually of set theory. What this meant is that, say, the merging of its parts into one another that constitutes a line[14] was replaced by a representation of it that involved *lots* of numbers, so many of them in fact that between any two of them there would be infinitely many and yet such that no two of them would merge into one another as we take the parts of a line to do—they would all remain as distinct from one another, as discrete, as numbers always are.

---

there lies a claim to absolute totality, as to a real idea, the very inadequacy of our faculty for estimating the magnitude of the things of the sensible world awakens the feeling of a supersensible faculty in us; and the use that the power of judgment naturally makes in behalf of the latter (feeling), though not the object of the senses, is absolutely great, while in contrast to it any other use is small. Hence it is the disposition of the mind resulting from a certain representation occupying the reflective judgment, but not the object, which is to be called sublime."

[12] So Kant can be seen as alternately adopting, like all of us, different logics at different times. His radical endorsement of Aristotle cited in Chap. 2 belongs in analytic logic, of course. But in (1987, pp. 146–149) I pointed to some dialectical steps he makes, when arguing from consistency to coexistence and then to mutual causal dependence. Now we see him also implicated in oceanic logic—even more distinctively implicated in it, given what his general philosophical outlook is, than in any of the others.

[13] As mentioned in footnote 20 in Chap. 6, the whole discussion conducted in Chap. 6 on Zeno's paradoxes revolves around the same issue that emerges here.

[14] As we are about to see, the whole problem I am addressing here could be reduced to what the parts of a line are. Consistently with footnote 12 above and the attending text, Kant claims that a line is not made of points; points are only limits of it and in general of space. "Space and time are *quanta continua*.... Space therefore consists only of spaces, time of times. Points and instants are only boundaries, i.e., mere places of their limitation" (1998, A169 B211).

9 Mathematics

The way continuity shows up in geometry is expressive of oceanic logic; "continuity," in fact, is a word often used by authors committed to that logic (e.g., as noted above, by Bergson) to bring out precisely the kind of merging referred to above. And, occasionally, that what is in question here be a logical, conceptual conflict has been recognized by major philosophers of mathematics, for example by Hermann Weyl (1987) when he states that "the conceptual world of mathematics is so foreign to what the intuitive continuum presents to us that the demand for coincidence between the two must be dismissed as absurd" (p. 108) and that "[w]hen our experience has turned into a real process in a real world and our phenomenal time has spread itself out over this world and assumed a cosmic dimension, we are not satisfied with replacing the continuum by the exact concept of the real number, in spite of the essential and undeniable inexactness arising from what is given" (p. 93). Effectively dismissing this important, early acknowledgment of inadequacy and even absurdity (Weyl's book was originally published in 1918), analytic mathematicians have soldiered on with their numerical representations, so I find it instructive to add the following remark: In 1900 Hilbert called the attention of mathematicians onto the continuum in a memorable way (at the Paris International Congress of Mathematicians), by making the assigning of a precise number to the continuum the very first of 23 unsolved problems that have kept the profession busy ever since. But what solution the problem has had did not relieve the underlying concern, for in 1964 Paul J. Cohen proved that the question is undecidable in Zermelo-Fraenkel's set theory.[15] While we may try to match our oceanic spatial intuitions analytically by throwing (literally!) countless numbers at them, we will never know what total number of them we are talking about. Oceanically, this is no big deal: it is only an indication that we are addressing the problem with the wrong tools. Analytically, it must be seen as one more sign of the inadequacy of the whole approach.

---

[15] See Cohen (1966).

Chapter 10
Texts

Abstract Meanings are interpreted, and what an interpretation amounts to depends on what logic we adopt. For analytic logic, ambiguity is a pathology to be avoided at all costs; therefore, a non-pathological carrier of meaning (a text) must have exactly one meaning. If the text was produced at time X, then what meaning it has at any other time must be the same as the meaning it had at X: the meaning of a text can only be its original meaning, and originalism is a necessity. Often, the original meaning is described in terms of the author's intentions, though that term must not be taken as metaphysically determinate since it is compatible with a wide variety of construals of what counts as an intention. However intentions might be construed, the uncovering of meaning will require the examination of data (about the author and her context) that might allow the interpreter to make a reasonable guess of what the original meaning was.

In dialectical logic, meaning evolves; hence, the interpretation of a text will be a function of the stage reached by the evolution of its meaning. The same text will carry different meanings at different times and in different cultures, sometimes opposite meanings from one time to another.

In oceanic logic, different readers (maybe at different times) will occupy different positions on one and the same spectrum, and their positions will be more reflective of their situations than of any objective feature of the text. The text itself, indeed, will remain a mystery to them, as they weave words around it primarily to signal and voice their limited perspectives on it.

Any occurrence can be interpreted; hence, any occurrence is a text. But, for someone to be participating in meaningful discourse, they cannot be participating in the production of meaning inadvertently, where "inadvertently" is to be read logically, not empirically—as not committed to a specific empirical way of understanding what it is to be inadvertent. Therefore, logically, there is no real distinction between expressing and interpreting a meaning (there cannot be one without the other), if by someone expressing a meaning we understand her to be engaged in meaningful discourse.

We are getting close to wrapping up, which we will do in the next chapter. As a preamble for that final move, it will be instructive to become officially aware of an essential connection that has been with us, implicitly, from the beginning and needs to be brought to the forefront now and inspected carefully. If a logic is a theory of the *logos*, that is, of meaningful discourse, then we must remind ourselves that the major thing that we do with meanings—besides expressing them—is to figure them out, fathom them, *interpret* them.[1] When we engage in a conversation, we may be trying to win an argument with our interlocutor; but, as a premise for running that argument, and possibly prevailing in it, it might be a good idea to understand what the interlocutor is saying (even if a successful strategy for defeating her might be to systematically misconstrue what we understood her to say). In a similar vein, we could not even begin a debate on whether or not a proof proceeds with cogent necessity, or on whether truth should escape error or redeem it, or negation should be taken as forbidding entrance into a territory or as an invitation to transcendence, if we were not satisfied that we had a correct interpretation of the proof, or of whatever truthful or errant statement we are considering, or of whether we take negation to be present.[2] If we agree to call any carrier of meaning a *text*, we can also agree that a large part of our concern with meaning has to do with interpreting texts, which, as it turns out, is anything but an innocent operation: because each logic conceives meaning differently, it also has a different conception of what it is to interpret a text. Since the science of interpreting texts is commonly called *hermeneutics*, we could also say that each logic comes with its own, distinct hermeneutics.

In analytic logic, as we have seen, ambiguity is an anomaly, and avoiding ambiguity is an obsession (whence the tendency, already noted, to multiply technical, artificially introduced and arbitrarily defined words and phrases). Therefore, every text must have a single correct interpretation; multiple readings are a pathology, to be blamed sometimes on the text itself ("a patchwork of various, incoherent drafts," "a hurried job, carried out against a deadline of publishing or perishing"), on its author ("he does not know what he wants to say"; "he is not making himself clear"; "he has not evolved a definite position yet"), or on the interpreters ("they are misunderstanding key terms"; "they are taking the text out of context"; "they are projecting their own meaning onto the text"). And, if a (non-pathological) text was produced 200 or 2000 years ago, its single meaning and its single correct interpretation must have remained analytically identical (identified by the same traits) through this entire time, as unchanged as the definition of *human* as *rational animal* is supposed to be. In line with what I said about the heuristics of a proof versus its logic, we may discover new nuances of a text's meaning long after it was composed, but only

---

[1] By the end of this chapter, it will become clear that there is a stronger relation between expressing and interpreting meanings than might appear now, and hence what exactly expressing a meaning is depends (as does interpreting it) on what theory of the *logos* we adopt.

[2] To illustrate this last point, consider the following: Leibniz thought that Descartes' ontological argument was missing a step, since before concluding to God's existence it had to establish that God's existence is possible (otherwise put, that God's definition, as the being who has all perfections, is not contradictory). And he believed that he could provide the missing step by arguing that perfections are all positive qualities and no contradiction can arise among positives. But (as Kant was later to point out; see my 1987, pp. 25–26) is "omnipotent," say, a positive quality? Isn't "X is omnipotent" equivalent to "There is *nothing* that X can*not* do"? So figuring out whether a negation is present is vital to passing a judgment on Leibniz's view.

because they were there all along and previous interpreters, somehow, had always missed them. Discovering a meaning after the fact, within this view, is like discovering a treasure that was once buried in the ground and had never before seen the light of day.

The significance of hermeneutics is greater to the extent that the texts it examines are of greater value to the community. So, not surprisingly, the science of interpretation has been most relevant and has had the most impact when applied to religious texts or to legal ones—especially, in the latter case, important ones like the US Constitution. In regard to both, an analytic logician will have to be an *originalist*, not as a matter of independent choice but as a necessity: whatever meaning the text has—and assuming that it does have a single, unambiguous meaning, the one that Aristotle finds indispensable for making sense—that meaning cannot be anything other than what the text meant *originally*, when first written. And, because the text was written by an author, there is a natural tendency to gloss its meaning, what the text means, as what its author *meant by it*: as what he *intended* to communicate to the readership by the act of writing it. This model works in a perfectly transparent way with religious texts, for those who take religion seriously: if a text is believed to be the word of God, then God will have unimpeded access to his intentions and the meaning of the text will be, uncontroversially and unproblematically, what he, by the issuing of it, wanted to communicate to the faithful. With humans, the metaphysics of intentions is a much more unsettled and unsettling arena, as it is ordinarily recognized that humans can have unconscious, repressed, and even conflicting intentions; hence, qualifying the meaning of a text as what its author meant by it is not a solution but a problem, not an answer but a question. We cannot expect the majority of legal (or biblical) scholars to have a high level of interest or expertise for such questions; therefore, their judgments on particular matters of interpretation might look quite fuzzy. But these are empirical issues (and complications); from a logical standpoint, their commitment to analytic logic will determine the landscape in which they operate—a landscape defined by the two basic points I just made: (1) the correct interpretation of (say) the US Constitution (of which there can only be one) must be today the same as it was for the founding fathers and (2) unveiling that correct interpretation is a process of discovery—of lining up facts about the time when the Constitution was written and by reference to them finding out what the original meaning was.

When we move from religious and legal texts to literary and philosophical ones, this conceptual framework is retained: (1) and (2) continue to apply. Which explains a practice endemic to interpreters of such texts: the unending search for yet unknown or unexamined documents that might provide the additional facts needed for unearthing a text's original meaning. If the meaning of *Don Quixote* or of the *Meditations* is today, and at all time, what it was for Cervantes or Descartes, then the perusal of a letter or of another minor work by the one or the other, which newly turned up in an archive, may hold the key to what that meaning was, and is, and forever will be. I talked about biblical and legal *scholars* above; here I note that there are, of course, scholars in the history of literature and of philosophy as well, and that they have a tendency to express disdain when someone ventures an interpretation that is not supported by a thorough investigation of, and consequent minute acquaintance with, all the details of the context of production of a literary or philosophical work (remember the complaint about taking a text out of context that I quoted earlier). Failing such investigation and acquaintance, there is a failure of scholarship, they superciliously observe, while themselves failing to realize that the very notion of scholarship is at stake here (as is their status as scholars): a scholar is someone who studies a text deeply (and scholarship is the quality she manifests when doing so); but studying is a form of engaging with the meaning of that text, and different conceptions of what meaning is will determine different forms of engagement, hence also different conceptions of what scholars and scholarship are. (Though it is to be expected—to go back full circle to the very beginning of this book—that, as long as people think that there is only one logic, they will think that there is only one way of construing the meaning of a text and engaging with it.)[3]

One additional remark on this topic is useful as a stepping stone toward our next theory of the *logos*. I mentioned the context of a text a couple of times; it must be pointed out that what counts as context is also a matter of controversy. The context of Descartes' *Meditations* might be taken to be his state of mind in the early 1640s, or his *oeuvre* up to that point, or his intellectual circle of friends and correspondents, or the collection of all his intellectual influences, or the entire philosophical, or cultural, Western tradition (an optional character that typically goes unmentioned by the stern defenders of placing a text in "its" context—a good example of using an improper definite description, and of dissembling by not admitting to it); but, even in the most inclusive understanding of the context which might be possible within analytic logic, the context will still be limited to what precedes, or is at most contemporaneous with, the writing of the *Meditations*. Because, once the *Meditations* were written, their one and only meaning had come to be, and nothing that happened later (none of the context that developed later) was going to be allowed to make a difference for it. It is a whole other story within dialectical logic—indeed, as we know, there it is a story: all meaning, including the meaning of any text, is an evolving structure, a narrative whose plot never stops unfolding, and turning against itself, and overcoming the internal contradictions it itself generates. Just as the meanings of "Socrates" and "Plato" have traveled a long way, and faced countless vicissitudes, in the centuries that separate us from Socrates and Plato, and will continue to do so in the centuries that follow us, so do the meanings of *Don Quixote* and the *Meditations*. Italo Calvino (1986) defined a classic as a work whose meaning is unceasingly added to, and enriched, by generations of readers. Empirically, that is a perceptive and illuminating thing to say; but for us, here, it is crucial to note that this empirical process is understood very differently by analytic and dialectical logics.

---

[3] If we regard the whole production of an author as a single text, the problem arises that there are often contradictions among the various statements the author makes at various times and in various works. In dialectical logic, which I will be exploring next, these contradictions are to be read as indications of the author's evolving meaning. In analytic logic, on the other hand (as I discuss in my 2000), there will be a tendency to apply the usual strategy of dividing and conquering: the "single" author will be split up into a family of distinct authors, each expressing his distinctive meaning—a precritical and a critical Kant, an earlier and a later Wittgenstein, etc.

The former must think of it as a straining and a perverting of the work's meaning, up to the point in which a whole new work is produced by the community of readers (and the time is perhaps ripe for a fresh look at the original work and its original meaning); the latter must think of it as constitutive of meaning itself, hence as being as much legitimate as it is inevitable.[4]

Aside from emotional commitments (to a person, to a cause, to a sports team), nothing makes people less disrespectful of one another—and more inclined to delegitimize one another—than having different logics, for in such circumstances the other guy seems unable to think straight, to give evidence of rationality: what he says is not (empirically) either right or wrong; it is straightforward nonsense. So let us consider a case in which both emotional attachments and different logics play a huge role. Emotions run wild on the issue of abortion, and that would be enough to explain why so many people have made it their mission in life to have *Roe v. Wade* repealed or sustained. But, if emotions were (counterfactually) not enough, different logics are also in play. The US Supreme Court's decision on this case, in 1973, was based on the following section of the Fourteenth Amendment: "No State shall make or enforce any law which shall abridge the privileges or immunities of citizens of the United States; nor shall any State deprive any person of life, liberty, or property, without due process of law." Based on this text, and more tentatively on the Ninth Amendment, which states "The enumeration in the Constitution, of certain rights, shall not be construed to deny or disparage others retained by the people,"[5] the Justices asserted a right to privacy that encompassed the right for a woman to choose whether or not to have an abortion; as a result, the Texas law banning abortion

---

[4]Though it is natural to account for a dialectically evolving meaning in terms of the history of how a text is understood (since, as was pointed out before, that is precisely what the *natural, immediate*, stance amounts to here), we must recall that dialectical development is not a chronological process but a logical one. Therefore, one could map the development within the text itself, without reference to attitudes taken toward it by empirical readers in different epochs; one could tell a (logical) story about how the text itself unfolds its evolving meaning. One might do this, on the surface, without espousing Hegel's optimistic, providential outlook: one might, for example, proceed as Jacques Derrida does (beginning with 1974 and 1978) and show the text to behave like a suicide bomber—to work out its meaning one step at a time until it blows itself (and its meaning) up—while implying (without ever quite admitting to it) that such a reading be a critical one (of the alleged consistency of the text). But proceeding that way requires adopting an uneasy, and ultimately incoherent combination of dialectical and analytic logic: conceiving of a meaning developmentally while also retaining an external point of view from which to observe that very development. If one is wholly committed to dialectical logic, as I pointed out in Chap. 3, optimism is not an empirical option but a logical necessity, since each phase of development comes with its own values—in this case, with the prevailing value of blowing texts up.

[5]The majority opinion of the Court (which was taking up a previous judgment, also favorable to the right to an abortion, by a Texas district court) stated that the "right of privacy, whether it be founded in the Fourteenth Amendment's concept of personal liberty and restrictions upon state action, as we feel it is, or, as the district court determined, in the Ninth Amendment's reservation of rights to the people, is broad enough to encompass a woman's decision whether or not to terminate her pregnancy." In a concurring opinion for a companion case, Justice William O. Douglas stated, less vaguely: "The Ninth Amendment obviously does not create federally enforceable rights."

which was the subject of their judgment was ruled unconstitutional (the only kind of judgment the Justices can proffer, as it is not part of their jurisdiction to make or to strike down laws, but only to decide on a law's consistency with the Constitution). Originalists like Antonin Scalia have claimed ever since that no such right to privacy, let alone a right to have an abortion, could be surmised as part of what the founding fathers meant when this text was published in 1789, or as part of the context in which that text was formulated; therefore, given that, whatever meaning the Fourteenth Amendment has today (or had in 1973), it must be the same for them as it had in 1789, it is impossible to interpret it as implying a right to privacy (or a right to have an abortion). And that, as we have seen, is what someone thinking, reasoning, and arguing analytically would *have to* believe and say—independently of what emotions he also feels. But that is also not the only way to think, reason, and argue.

Someone committed to dialectical logic would see a radically different picture. For her, the meaning of the Fourteenth Amendment has changed considerably between 1789 and 1973 and has occasionally been turned on its head. It is quite likely, for example, that with their mention of "citizens" the founding fathers, and their context of operation, could only have intended (white) males (however intentions are construed) and that, if any thought of a right to privacy ever crossed their minds, it would be the right that an adult (white) male was recognized to have, back then, to enforce whatever rules and discipline he saw fit to impose within his household. In 1973 the meaning of "citizen" had gone through a momentous transformation and so had, consequently, any reference to their privileges and immunities or to what it means to deprive a person of liberty or privacy. Sexual freedom, say, was not contemplated as part of the meaning of liberty or privacy, which is why many states had, for a long time, laws about what people could or could not do in their own bedrooms that were not declared unconstitutional.[6] Therefore, dialectically understood, the Fourteenth Amendment not only can but must be interpreted at variance with anything the founding fathers, or their context, might have intended: it must be brought up to date, made responsive to its own history and to the current stage of development of all of its key terms. And it will be just as inescapable an outcome that people bringing such radically different pictures to the task of interpreting the meaning of the Fourteenth (or Ninth) Amendment will not be able (even if they were open to it, emotionally speaking) to appreciate the grain of sense that might be contained in their opponents' interpretations: that they will totally disqualify it as something that does not even belong in the range of respectable discourse (or reading or scholarship).

One last logical lesson can be learned from this example. It takes dialectical logic to find a meaning in the Fourteenth Amendment that is supportive of the right to an abortion, but this is not to say that dialectical logic itself is necessarily and unqualifiedly supportive of such a right. For the problem is: there is nothing definitive in dialectical logic; there are no conclusions. The meanings of "liberty," of "citizen," and of all other key terms in the Fourteenth Amendment did not cease to

---

[6] The Supreme Court struck down sodomy laws as unconstitutional only in 2003, in *Lawrence v. Texas* (with Scalia and two others dissenting).

evolve after 1973, and it is possible that their later evolution has once more changed them (or will change them) in ways that are inimical to the right to an abortion. The prospect does not look implausible, given the current intellectual state of the nation. So one should be wary of pro-choice people defending at the same time the legitimacy of an evolving meaning of "liberty" and like terms up to 1973 and the illegitimacy of renegotiating, now or in the future, what they regard as a definitive achievement. More generally, one should keep in mind a crucial cautionary note: a logic will not decide a social, political, or otherwise empirical issue, though it will decide whether, in discussing that issue, people are in a position to understand one another (assuming they want to).

In light of the above, it is worth revisiting biblical interpretation. God may well be regarded as the epitome of someone who is absolutely clear about what he means, but that does not imply that the readers of texts which are thought to be the word of God are absolutely clear about what those texts mean, empirically or even logically—that they are even clear about what God's logic is. Consider a case that had great historical significance. In *Joshua* 10: 12–14 we read:

> Then Joshua spoke to the Lord in the day when the Lord delivered up the Amorites before the sons of Israel, and he said in the sight of Israel, "O sun, stand still at Gibeon, and O moon in the valley of Aijalon." So the sun stood still, and the moon stopped, until the nation avenged themselves of their enemies. Is it not written in the book of Jashar? And the sun stopped in the middle of the sky and did not hasten to go *down* for about a whole day. There was no day like that before it or after it, when the Lord listened to the voice of a man; for the Lord fought for Israel. (*New American Standard Bible*)

How is this text to be read? Whatever does God (who, supposedly, inspired it) mean? For someone committed to analytic logic, he can only mean what he meant *then*, as there is only one (analytically) identical meaning of what he says; and what he meant then could only be that the sun should (and did) stop its course in the sky, hence that the sun, independently of this miraculous interruption, was moving in the relevant sense across the sky, around the earth. Whence it follows that the Aristotelian-Ptolemaic model of the universe, with the earth standing still at the center and the sun, the moon, and all the planets revolving around it, is hereby validated by biblical authority; for the believer, it must be the plain, indubitable truth, since thinking otherwise would require denying the very word of God.

One could also, however, conceive of God thinking differently. Bruno (1995) says:

> It is certainly appropriate, when one seeks to direct history and to give laws, to speak according to the common understanding, and not to be concerned with indifferent matters.... So if the Sage, instead of saying, "The sun riseth and goeth down, turned toward the south and boweth to the north wind," had said, "The earth turns round to the east, leaving behind the sun which sets, bows to the two tropics, that of Cancer to the south and Capricorn to the north wind," his listeners would have stopped to think: "What, does he say that the earth moves? What kind of fables are these?" In the end, they would have accounted him a madman and he really would have been a madman. (p. 178)

If God is the supreme sage, he will know how to adapt his words to the mindset of the readers; hence the proper way of understanding him, in any epoch, would be according to the state reached in that epoch by the evolution of the meanings of all of his terms. In the sixteenth century, after Copernicus, it would have been silly to read God as still committed to the geocentric model (and so would it be today).

We know that this protodialectical understanding of God's word did not turn out well for Bruno; nor did it for Galileo, who (despite an early, and probably hurried, acceptance of the *Dialogue concerning the Two Chief World Systems* by the Holy See) was eventually tried for similar statements, found guilty, and was not rehabilitated until 1992. I could, of course, regret and indict such occurrences; but this is not the place for statements of the sort. It is the place to point out, rather, that, even when it comes to religious texts, originalism (or, as would be called in this kind of conversation, fundamentalism) is not a destiny, not unless you espouse analytic logic. Dialectical logic is always there as an alternative, to remind you that, if God's own meaning never changes, how God's meaning is to be understood (and even how he chooses to phrase it at any one time) may well be a function of what audience he is addressing—because "never changing" depends here on a dialectical notion of identity. Within this framework, biblical hermeneutics would then be as much a search for an evolving meaning as its legal or literary counterpart.

What attitude does oceanic logic bring to the reading and interpretation of a text? One that attenuates, to the point of making it disappear, the oppositional character of any different readings of the text. Take the Fourteenth Amendment once again and the liberty invoked in it, and the vexed right to an abortion that (according to some) the invocation of such liberty supports. From an analytic point of view, there is going to be a stark contrast between two applications of the same liberty—the liberty of the woman to be free of child and the liberty of the child to be free to live—and there is going to be a zero-sum game to be played between the two. From a dialectical point of view, there is going to be a narrative accounting for how the concept of liberty develops from a stage in which the child's (and maybe the species' or society's) interests dominate to another one in which the woman's do (possibly to be followed by yet another in which the tide turns back). From an oceanic point of view, a gradual argument, sorites-like, is going to be worked out in order to establish that the two, allegedly competing, interests are actually one and the same. How far, and specifically how long, can a pregnant woman go in distancing her identity and welfare from the identity and welfare of the prospective life she carries? Four weeks? Three months? A couple of years? How far can a prospective child go in distancing his identity and welfare from the identity and welfare of the woman who carries him? Can the identity and welfare of the one be defined on the basis of sheer indifference to the identity and welfare of the other?

It will be useful to reiterate that the above is not encouraging a specific position within the debate on abortion. Such a position would be an empirical one, to be located in the province of someone's empirically held social, political, or moral convictions. That one is committed to analytic logic does not determine which of two (or several) contrary positions she will take in the zero-sum game defined by this logic; that one is committed to dialectical logic does not determine which narrative she will tell or which chapter of the narrative she will regard as her own— as her own time comprehended in thoughts. That one is committed to oceanic logic does not determine where she will locate herself in a spectrum of infinite graduality, where each location insensibly melts into others and yet one is still placed at one location or another, looking at things from a particular point of view. But the commitment to a logic does determine how one sees other positions and oneself as possibly assuming them: as the difference between right and wrong, and as oneself being lost in error; or as a more primitive, more immature stage of development, and as oneself being driven by a nostalgic fervor toward the innocent blindness of one's own youth; or as a neighbor who has simply chosen a slightly (or not so slightly) different point of balance on the same spectrum and as oneself sliding into her point and acknowledging her to be addressing the same constellation of worries—to be manifesting the same baldness or non-baldness, with only one more, or one fewer, hair on her scalp. Undeniably, this last stance is more modest than the other two, more accepting of the alternatives, more open (we know) to compromises that are not just cases of selling out. And that is because (as we also know) it is a stance that comes with a recognition of its own limits to transparency: because it thinks of itself as weaving a fabric of ultimately inconclusive words over a reality that is alien to words—that will not be conquered by them. I will now articulate this attitude, with specific reference to texts, by using a Hegelian distinction.

In (1984/1988) and elsewhere, Hegel distinguishes between a secret and a mystery.[7] A secret is something that could well be understood, except that it is hidden, kept away from view; a mystery is something that is hard to figure out, though it is under everyone's eyes. Hegel's own logic, he believes, has no room for either secrets or mysteries: everything in it is perfectly clear, including the meaning of God's words and the meaning of any human text. Whether or not we admire his *hybris* (but without buying into it), we can use his terminology to characterize the other two logics' stances with respect to texts, divine or otherwise. Analytic logic proceeds in the wake of the secret: the one and only meaning of a text, especially an important one, *could* be made perfectly clear but often is not, because often it is hidden "between the lines." So the interpreter's job is uncovering it: looking carefully in all the nooks and crannies for the map to the treasure, for the key that will unlock the door, for the combination that will crack the safe. Oceanic logic, on the other hand, proceeds in the wake of the mystery: the text, like God and like all reality, will inevitably escape us; the best we can do is put words around it, which more than anything else will reflect *our* situation—our needs, our desires, our goals, what gives us pleasure and what pains us. And, though there is no question that, at any one time, we will identify with that situation of ours, it is also the case that, insofar as we see ourselves that way, we will disagree with those in different situations, and even antagonize them, but not dismiss them or demean them, since we know that it would only take one hair more or less, one slightly lighter or darker shade of color, to come to see things, and find the same meaning in a text, as they do.

One final remark. I started this chapter focusing on one main thing we do with meanings—interpreting them—and distinguishing it from another main thing we

---

[7] "[I]n the Christian religion there is no longer any secret—a mystery, certainly, but not in the sense that it is not known" (1984/1988 I, p. 382).

seem to do with them: expressing them. Now I need to point to the delusiveness of that distinction. It is certainly possible that we "express" a meaning by an utterance or a gesture in the same way as the sky does when it means rain for a peasant or an infant does when it means hunger by its restlessness. Then, however, we do not belong in the meaningful *discourse* that includes our utterance or gesture; for us to enter the *logos*, our behavior must be meaningful to us as well—that is, we must also be spectators of it and we must read a meaning in it as much as anyone else does, though not necessarily better than anyone else and though not necessarily the same meaning, or even according to the same logic, as anyone else. We cannot, in other words, express meaning inadvertently, as long as this adverb is properly understood: as signaling a logical, not an empirical condition. Much as it is possible that there be intentions that are unconscious and unknown to the very person having them and may be better known to others, hence to require that some behavior be intentional (e.g., behavior we may be legitimately regarded as responsible for) is only to phrase a logical condition that can be satisfied by any number of different empirical circumstances (e.g., by behavior that attains goals the agent can be independently recognized as having), so saying here that an agent cannot express meaning inadvertently is compatible with all sorts of impulsive or unconscious performances, as long as something in the agent's behavior can be construed as manifesting the awareness required by the logical condition. With that qualification, expressing a meaning is seen to be no different than interpreting it, at least in the sense that any case of the one is also a case of the other. From which it follows, of course, that expressing a meaning is open to the same choice among different logics as interpreting it is. What meaning I express with any of my behavior (e.g., in producing this book) can be read as a once-and-for-all affair, decided at the moment of conception (or of writing, or of signing off the final draft to the press), or as a narrative that will continue to evolve long after my physical death, or as a precarious compromise always threatening to slide into a neighboring position.

Chapter 11
The Play of Logics

Abstract The selective attitude expressed by analytic *formal* logic, which disregards much of the content of what we say to concentrate on what it calls the form of discourse (and is nothing other than a portion of the content), can be, and has been, generalized to an attitude according to which what has logical, conceptual dignity does not belong to the ordinary, spatiotemporal world but subsists in its own transcendent dimension, and ordinary objects and experiences only provide inessential, imperfect examples of logical structures. For dialectical logic, on the other hand, the concept inhabits the spatiotemporal world, that world is the concept incarnate, and every minute detail of the world illuminates the concept's logical vicissitudes. For oceanic logic, the world is an unresponsive, mute mass that is closed to any attempt at deciphering its code; it is what Wittgenstein called the mystical.

Since a theory of the *logos* is an all-inclusive doctrine, among the countless things it is supposed to account for are the competing theories. For analytic logic, there can be only one definition of the word "logic" (ambiguity is an anomaly there); so alternative logics would be nonsensical (though there may be competing theories that give distinct formulations of *one and the same logic*, as in the proliferation of logics we have witnessed in the last several decades). Dialectical logic appropriates the other logics as early phases of itself: the contraries that analytic logic regards as definitive chasms in the structure of the world are instrumental in bringing out the conflicts that will be dialectically transcended, and Spinoza's oceanic view is understood as a rudimentary (immediate) manifestation of the idea. Oceanic logic regards its rivals as different, partial points of view on the same (mysterious) logical reality.

Finally, how shall we behave in the presence of alternative, and opposed, theories of thinking, reasoning, and arguing? If language is seen as less a means of communication than an arena of play, then the answer is: playfully. In our everyday conversations, we constantly shift from one logic to another; and all such shifts empower us, by making us familiar with all the diversity that is reflected in them.

In Chap. 2 I mentioned the Kantian thesis that our language determines our world and said that here I will leave it aside. But, without bringing up such a radical view, we might find it appropriate to endorse a weaker, far less controversial one: that how we understand (meaningful) language—which is to say what logic we use—deeply influences how we *think* of the world. This topic has already surfaced in the above, and it is time to consider it in more detail.

The crucial point here, for analytic logic, is that it (and it only) can be formalized: some of the content of what we say can be abstracted from, let go, judged irrelevant to logical concerns, and the rest of it (qualified as form) focused on as logic's proper subject matter. This selective attitude grounds the divide between concepts and objects (as well as the debate about what concepts are: the great debate on the nature of universals) and the classical distinction between pure and empirical knowledge: between what can be attained by reflecting in an armchair on the logical relations constituting the very *órganon* of knowledge and what can only be discovered by getting up from the chair, gathering (factual) data (*givens*—something that is not accessible to pure thought, that must be received), and applying the *órganon* to accounting for them. Coming up with such *scientific* accounts is, of course, a troubled business, torn between two equally unpalatable horns: between using concepts in merely *classifying* the data (which is a lot of what Aristotle does, e.g., in the biological works) and venturing the claim that what the *órganon* shows to be logically necessary must also be factually true—thus implicitly assuming the world to be reasonable and dangerously evoking the radical thesis we are not going to get into here.[1]

What we will get into is how, within this outlook, it is hard to confer any intellectual dignity on the data. There will be few who are ready to go, in debasing them, all the way with Plato: to claim that logic is not just an *órganon*—a tool, that is—but a map to the whole of reality, and what falls outside it does not even exist; it is a phantom of our brain. But even those others who show more respect for the data, and judge them decisive for establishing (or, rather, confirming) the correctness of a

---

[1] Both Kant and Hegel, despite their (somewhat) different logics, recognize that the assumption of an agreement between natural occurrences and rational considerations plays a major heuristic role within scientific endeavors. See, for example, the following passages: "It is in fact indispensable for us to subject nature to the concept of an intention if we would even merely conduct research among its organized products by means of continued observation; and this concept is thus already an absolutely necessary maxim for the use of our reason in experience. It is obvious that once we have adopted such a guideline for studying nature and found it to be reliable we must also at least attempt to apply this maxim of the power of judgment to the whole of nature, since by means of it we have been able to discover many laws of nature which, given the limitations of our insights into the inner mechanisms of nature, would otherwise remain hidden from us" (Kant 2000, p. 269). "While at first it is only dimly aware of its presence in the actual world, or only knows quite simply that this world is its own, ... [reason] strides forward in this belief to a general appropriation of its own assured possessions, and plants the symbol of its sovereignty on every height and in every depth. But this superficial 'mine' is not its ultimate interest; the joy of this general appropriation finds still in its possessions the alien 'other' which abstract reason does not contain within itself. Reason is dimly aware of itself as a profounder essence than the pure 'I' *is*, and must demand that difference, that being, in its manifold variety, become its very own, that it behold itself as the *actual* world and find itself present as a shape and thing. But even if reason digs into the very entrails of things and opens every vein in them so that it may gush forth to meet itself, it will not attain this joy; it must have completed itself inwardly before it can experience the consummation of itself" (Hegel 1977b, p. 146; translation modified). It is hard, on the other hand, to find scientists who are even aware of this possible presupposition of their entire work.

scientific conjecture, will probably agree with Kant that there is as much science in a discipline as there is mathematics (i.e., pure, rational knowledge) in it[2] and hence will think that all scientific work is done in drawing the mathematical consequences of the conjecture, and then, when we finally get to a point where some of these consequences can be compared with the data (say, with what actually happens during a solar eclipse), we run that comparison to test whether the science we have been doing is actually *of this world* (and, if it is not, Einstein was reported as saying, so much the worse for the test itself, and by implication for the world[3]). To be sure, designing and running experiments is a much more challenging enterprise than would be suggested by this simple addition of a demonstrative, but remember: I am not focusing here on the practice of science (or anything else) per se, but on the way we are constrained to think of that practice by what logic we adopt. And, within the analytic way of thinking, we have not gone much further, in explaining how science constructs its accounts of the world, than the already-mentioned hypothetico-deductive model: you conjecture, you (logically) deduce, and eventually you test. In this operation, the world is never taken to have anything to contribute to thinking and reasoning themselves: it is only a repertory of examples, positive or negative, of what issued from thinking and reasoning, a yes-or-no switch that shows thinking and reasoning to be either right or wrong.

It is quite otherwise in dialectical logic. Here the concept penetrates the world, to its most intimate recesses; the world is nothing other than concept. Using a Christian metaphor that Hegel cherished,[4] the world is the concept incarnate. An abstract definition of *human* (say, as *rational animal*) can be at best a largely inarticulate starting point for a conceptual account of humans. What needs to be done next, and forever, is to follow up all the vicissitudes of *human* in the world and connect them into a single narrative that makes every one of them appear necessitated, in the specific sense of necessity that circulates in this logic. The narrative will have to incorporate the fact that *human* breaks down into *male* and *female*, and into different groups that (at least at some stage of the narrative) are taken to constitute different races, and that humans form families and aggregate in societies and give themselves laws and then societies face one another in bloody wars and make peace treaties and commercial agreements and on and on for everything that is ever attributed to humans. All of this, it must be emphasized, will not be something that just *happens* to humans: to beings identified, once and for all, by the clause *rational animal* or by some other combination of genus and differentia. It will not be something that is only factually true of those beings that are necessarily, essentially, rational animals.

---

[2] See Kant (2004), p. 6.

[3] See Armstrong (2009): "when asked what would happen if his theories were not vindicated in the laboratory, ... [Einstein] retorted, 'So much the worse for the experiments; the theory is right!'" (p. 267).

[4] See his (1977a), pp. 180–181: "Religion offers a possible reconciliation with nature viewed as finite and particular. The original possibility of this reconciliation lies in the original image of God on the subjective side; its actuality, the objective side lies in God's eternal Incarnation in man, and the identity of the possibility with the actuality through the spirit is the union of the subjective side with God made man."

It will be as much part of their essence as rationality and animality are supposed to be; indeed, it will as much articulate their rationality and their animality as it does their humanity. And, as more and more elements are added to the necessary constitution of humans, it will become clearer that all those elements (war, peace, society, treaty, etc.) compose a single integrated structure, pervaded by the same conceptual necessity, and Hegel's monistic drive will come to the fore: we are only ever talking about the same thing, about the spiritual unity of all that is, about how there is not *this* concept and *that* concept and *that other* concept, but only *the concept*—the entire world logically redeemed.

Any number of projects that have shaped the culture of the last century or so can only be seen to make sense in this framework. Why would sophisticated intellectuals like Walter Benjamin (1969), Theodor Adorno (1974), or Roland Barthes (1983) spend so much of their valuable time investigating reproduction in contemporary art, or the astrological column of the *Los Angeles Times*, or the world of fashion? Because they had run out of other, more important things to say, and decided to invest their considerable brainpower in providing highbrow journalism to rich, largely illiterate bourgeois: to categorize in a becoming manner the objects of the latter's obsessions and sins? This is what we would be forced to answer if we went Aristotelian about it; but we can go differently. We can take our lead from Hegel instead, and then we will *have to* think that zodiacal ascendants and glamorous runways, limited edition prints and Xerox copies, and (why not?) the Super Bowl and the Hearst Castle are but conceptual twists and turns, as revealing of what spirit has been up to as any academic text.

So here we have a clear-cut contrast: for analytic logic, the world is a tiresome redundancy, as all that matters occurs away from it, in ethereal, tenuous conceptual heaven; for dialectical logic, the world is shot conceptually through and through, so much so that, if you want to learn more about the concept, make it what Hegel would call more concrete, you must interrogate the world. How about oceanic logic? What does the world look like when all distinctions we might want to make are judged delusional—so many cases of much ado about nothing?

It looks as one would expect: as an unresponsive, mute mass that is impervious to any attempt at deciphering its code, that remains unfazed before any description one might proffer of it. I have cited Bergson a few times as someone committed to oceanic logic; here is what he says about the most radical contrast ever posited by philosophy—the one between matter and mind.

> Thus, in renouncing the factitious unity which the understanding imposes on nature from outside, we shall perhaps find its true, inward and living unity. For the effort we make to transcend the pure understanding introduces us into that more vast something out of which our understanding is cut, and from which it has detached itself. And, as matter is determined by intelligence, as there is between them an evident agreement, we cannot make the genesis of the one without making the genesis of the other. An identical process must have cut out matter and the intellect, at the same time, from a stuff that contained both. (1911c, p. 210)

The way the "cutting" operates is, I said, by projecting onto the same undifferentiated stuff a particular "point of view," which is conceived by Bergson as determined by (and expressive of) practical utility: "Intelligence, in its natural state, aims

11 The Play of Logics 145

at a practically useful end" (ibid., pp. 163–164). But this whole enterprise must be exploded (perhaps by paradoxes?): "You must take things by storm: you must thrust intelligence outside itself by an act of will" (ibid., p. 204). And some other form of relating to reality must be tried: something that is not as transparent as intelligence promises to be, not as able to be communicated in words, but is more in touch with the unspoken, unspeaking, enigmatic nature of that reality.

> But we can conceive an inquiry turned in the same direction as art, which would take life *in general* for its object, just as physical science, in following to the end the direction pointed out by external perception, prolongs the individual facts into general laws. No doubt this philosophy will never obtain a knowledge of its object comparable to that which science has of its own. Intelligence remains the luminous nucleus around which instinct, even enlarged and purified into intuition, forms only a vague nebulosity. But, in default of knowledge properly so called, reserved to pure intelligence, intuition may enable us to grasp what it is that intelligence fails to give us, and indicate the means of supplementing it. (ibid., p. 187)

Where the very contrast that was just brought out between intuition and intelligence must be resolved, by the usual slippery-slope technique, into undifferentiated unity (*the same* undifferentiated unity as always):

> we traced a line of demarcation between the inorganic and the organized, but we pointed out that the division of unorganized matter into separate bodies is relative to our senses and to our intellect, and that matter, looked at as an undivided whole, must be a flux rather than a thing.... [W]e have shown ... that the same opposition is found again between instinct and intelligence, the one turned to certain determinations of life, the other molded on the configuration of matter. But instinct and intelligence ... stand out from the same background, which, for lack of a better name, we may call consciousness in general, and which must be coextensive with universal life. (ibid., p. 196)

In a later French author, Gilles Deleuze, a dual ontology is given. At a deeper level, there are bodies, which exist in a timeless present, to which nothing ever happens. They are causes, but not in the sense of antecedents necessitating (temporal) consequents: in the sense of being the root, the foundation of all antecedents and consequents—of everything that can be expressed in language.

> First, there are bodies with their tensions, physical qualities, actions and passions, and the corresponding "states of affairs." These states of affairs, actions and passions, are determined by the mixtures of bodies.... There are no causes *and* effects among bodies.... Second, all bodies are causes in relation to each other, and causes for each other—but causes of what? They are causes of certain things of an entirely different nature. These *effects* are not bodies, but, properly speaking, "incorporeal" entities. They are not physical qualities and properties, but rather logical or dialectical attributes. They are not things or facts, but events. We can not say that they exist, but rather that they subsist or inhere (having this minimum of being which is appropriate to that which is not a thing, a nonexisting entity). They are not substantives or adjectives but verbs. They are neither agents nor patients, but results of actions and passions. They are "impassive" entities—impassive results. (1990, pp. 4–5)

How is this different from a God who is the timeless, eternally present foundation of the universe: of something that does not properly *exist*, insofar as it owes all of its being to God, as it is but an excrescence on his surface, a wavy shape appearing on it, and nonetheless will forever try to verbalize him (that is part of the shape the wave takes), to phrase him in the meaningful currency of the *logos*—to call him omnipotent, omniscient, merciful, supremely just, or whatever else fits its diverse, fleeting interests? And how is it different from the mystical in Wittgenstein's *Tractatus*?

> 6.44 It is not *how* things are in the world that is mystical, but *that* it exists....
> 6.5 When the answer cannot be put into words, neither can the question be put into words....
> 6.52 We feel that even when *all possible* scientific questions have been answered, the problems of life remain completely untouched. Of course there are then no questions left, and this itself is the answer.
> 6.521 The solution of the problem of life is seen in the vanishing of the problem....
> 6.522 There are, indeed, things that cannot be put into words. They *make themselves manifest*. They are what is mystical....
> 7 What we cannot speak about we must pass over in silence. (pp. 149–151)

In analytic and dialectical logics, concepts (or the concept) capture either a limited part or the whole of the world. In oceanic logic, they look at the world askance, as if to something foreign. They know that they do not belong there, that no articulation they can submit will capture what is intrinsically inarticulate; hence, they might derive a healthy dose of humility, an attitude of surviving in the exile condition of a vale of tears by recognizing their common predicament and negotiating compromises with other, equally unenlightened exiles.

These being the logics' different postures, we can reconsider and better understand their favored conversational strategies, which were mentioned in Chap. 1 and then played out in the various interludes. The analytic selective stance inclines toward a take-no-prisoner attitude: *my* selective account of the conceptual essence of the world can only be vindicated if it refutes *yours*—as long as yours is a contrary one. The dialectical global stance, on the other hand, must keep in mind that the other guy's contrary view is itself part of the whole world which must be made to display a thoroughly conceptual structure; hence that view, too, must not be straightforwardly, naively countered but, instead, made perfectly good sense of—one must be made to understand how that *error* had to come to pass; how it was even good that it did. This attitude, to anticipate a theme that will be discussed shortly, can be illustrated by reference to Aristotle himself. In Book VII of the *Nicomachean Ethics*, the Philosopher is opposing those who claim that physical pleasures are the greatest good, and he adds the following remark:

> Since we should state not only the truth, but also the cause of error—for this contributes towards producing conviction, since when a reasonable explanation is given of why the false view appears true, this tends to produce belief in the true view—therefore we must state why the bodily pleasures appear the more worthy of choice. (1154a)

The one who reasons according to oceanic logic, finally, will find negotiation, compromise, and eventual agreement (possibly to disagree) to be most natural: she will see herself and all other interlocutors as exploring distinct, equally myopic and parochial, but also equally profitable outlooks on one and the same unknowable reality—will see a joining of all these paltry forces as the best way of navigating an unruly ocean.

As a particular, extreme case of a conversational strategy, consider the following: A theory of the *logos* is an all-inclusive doctrine—in the Kuhnian terminology I introduced in Chap. 2, it is a paradigm. Therefore, among the countless matters it must account for, it must also have something to say about its rivals: about the fact that there are people out there who do not think and reason according to its canon. We know from Chap. 7 how Aristotle (officially) dealt with alternative ways of thinking: within his antagonistic logical habitat, they amounted to *non*-thinking. Whoever states that *human* can turn into *nonhuman*, or *mortal* into *immortal*, or a man into a trireme, is giving up on meaning altogether: he can babble but cannot speak, and in any case cannot engage or be engaged in a reasonable conversation. Presumably, the same is true for those who state that *pink* and *red*, or *bald* and *nonbald*, are different ways of referring to the very same thing, which is equally both, depending on how you look at it. Anomalies like the sorites, or like the identity in difference of space-time continuants, will be regarded as problems to be solved; and, if we cannot solve them, it will be our fault for not being smart enough. There must be as much a single definition for the word "logic" as there is for "human," by genus and differentia; and, by the definition (implicitly) provided by Aristotle, those other theories have nothing to do with logic—in fact, they are not even theories, as they make no sense.

Dialectical logic has, with respect to the other logics, the same attitude it has toward everything: they are just right, at the stage at which they are located. In Chap. 4 I mentioned the fact that Hegel appropriated Spinoza's substance and treated it as an early manifestation of the idea; a more detailed version of that remark will give us the unity promulgated by oceanic logic as an early, *immediate* intimation of the unity of spirit, and the inarticulate character of that unity as one that (despite oceanic logic's protestations) will inevitably, necessarily be resolved into the seamless, infinitely transparent narrative of spiritual development. Dialectical logic will thus be *the truth of* oceanic logic as a tree is the truth of a relatively unstructured seed.

The same diagnosis will apply to analytic logic, and here Hegel, belaboring a point suggested before, will call Aristotle himself as a witness. In his (1995), after lavishing enormous praise on Aristotle ("[Aristotle] was one of the richest and deepest of all the scientific geniuses that have as yet appeared—a man whose like no later age has ever yet produced," II, p. 117), Hegel rejects the traditional reading of the Aristotelian corpus (and practice) as consisting of a logical *órganon* to be prefaced to science (and philosophy) proper and pictures it as developing a logic that gets progressively richer and deeper than what is offered in those early treatises—for all intents and purposes, a *dialectical* logic:

> Aristotle's logic has treated the general theory of conclusions in the main very accurately, but they do not by any means constitute the universal form of truth; in his metaphysics, physics, psychology, etc., Aristotle has not formed conclusions, but thought the concept in and for itself. (ibid., p. 217)

> Aristotle is thus the originator of the logic of the understanding; its forms only concern the relationship of finite to finite, and in them the truth cannot be grasped. But it must be

> remarked that Aristotle's philosophy is not by any means founded on this relationship of the understanding; thus it must not be thought that it is in accordance with these syllogisms that Aristotle has thought. If Aristotle did so, he would not be the speculative philosopher that we have recognized him to be; none of his propositions could have been laid down, and he could not have made any step forward, if he had kept to the forms of this ordinary logic. Like the whole of Aristotle's philosophy, his logic really requires recasting, so that all his determinations should be brought into a necessary systematic whole—not a systematic whole which is correctly divided into its parts, and in which no part is forgotten, all being set forth in their proper order, but one in which there is one living organic whole, in which each part is held to be a part, and the whole alone as such is true. (ibid., p. 223)

The analytic theory of discourse (and thought)—a theory of contrariety, which stresses the unbridgeable chasm between contraries—is (from a dialectical standpoint) what is needed to make sure that contraries, or contradictories, are not seen as genteel differences: that the intense drama of their opposition is properly emphasized, to make for adequate logical fulfillment, and ecstasy, when that opposition is overcome. Thus, dialectical logic gets unity from one of its rivals and the most extreme form of negation and exclusion from the other, and can happily see itself as transcending *their* opposition and as being the truth of both.

Oceanic logic will judge the confrontational stance of one of its rivals and the condescending stance of the other as the playing out of the limitedness of all points of view: the destruction of opposites and the incorporation of them will be considered distinct, but ultimately equivalent, strategies in the enterprise of defending one's correctness—of not accepting that all correctness be reduced to error. These strategies will be useful, of course, as they refine and strengthen each point of view and make the error it represents more of a resource in surveying the boundaries and failures of language, in voicing how much is being missed. To quote from Bergson one last time (and to bring up continuity once more):

> The real whole might well be, we conceive, an indivisible continuity. The systems we cut out within it would, properly speaking, not then be *parts* at all; they would be *partial views* of the whole. (1911c, p. 32)

So much for how each logic will deal with the others. The final question to be addressed in this chapter, and in this book, is: how should we regard the plurality of logics as a whole and behave in the face of it? Is it not confusing that there be not just alternative organized systems of thought that conceive of the world in disparate ways but also alternative organized systems that conceive of *what it is to think* in disparate ways? How can we do research together, or even carry out some kind of dialogue with one another, if the very texture of that dialogue, or the very language in which that research is formulated, is understood differently by different participants in it—or even by the same participant at different times? Shouldn't we make up our minds and decide which of these theories of the *logos* is to be adopted and maintained throughout? Shouldn't we be doing all we can to avoid the Babel that would issue otherwise: the confrontation of mutually incomprehensible tongues?

No answer to any of these questions is theoretically innocent, though a pretense of theoretical innocence (before it is exposed—before someone shouts that the king has no clothes) is a requirement for a definitive "proof" of the correctness of one's view. Answers to them can only be provided by building on some theory of what the nature and role of language, of research, and of knowledge (among other things) are. My answers build on my theory of such matters, which I articulated and defended elsewhere but I am going to simply state here.[5]

Language is not a means of communication: we can, and most often do, "say" everything we need to say to each other without using words. Language is, rather, a vehicle of play, an immensely supple medium that can model virtually any situation, and then those models can be stretched and twisted, in the course of doing "research," much as a toddler stretches and twists objects of common use—if she can lay her hands on them. In the grand scheme of things, this play, like all play, has adaptive value: we get to explore possibilities and vicariously live in them, and sometimes those possibilities go viral and become the realities of tomorrow. We come to trust what is reflected in a lens more than our own eyes; we find ourselves surrounded by Turing machines. But this is not to say that play should be performed because of its adaptive value: play is a gratuitous activity, the best example of an Aristotelian *prāxis*[6]; whoever plays with anything other than play in mind, with some ulterior motive, has violated its spirit—and probably will not draw the dividends of its adaptiveness.

All play, including play in language, requires free space and time. The play of a screw in its hole is my inspiration here: if its fit were snug, it could not move. Generalizing on this modest, precious example, if I find myself fitting perfectly with my environment, my opportunities for play are gone. In order to play, I need a certain amount of dissonance, of heterogeneity, of distraction and bewilderment. One tyrant spells servitude; two opposing tyrants may spell some degree of liberation. Therefore, I need dissonance in my most recurrent and widespread arena for play: when I play in, and with, language. If a logic is a structured manner of understanding what language is—how it differs from mere sounds and inkblots—then a plurality of such manners gives me a maximum of freedom, the amplest room for exploiting the playful vocation of language: for playing not only with what language lets us express, but with what language itself is. If one values, as I do, play and freedom, then the plurality of logics is a precious gift: one that it would be anathema to try to regiment away.

In my everyday life, I find myself (and find others) shifting constantly between logics. If I hear a corrupt or inefficient politician trying to defend something he did by obliquely referring to the slippery slope between honesty and dishonesty, or success and disaster, or by describing where he (and what he did) comes from, I want to crucify him to the strictest, most literal analytic understanding of his words and actions. If I have a talk with my children, I tend to see their lives, and their joys and

---

[5] See primarily my (1997) and (2013).

[6] "[P]ractical wisdom [*phrónesis*] cannot be knowledge nor art; not knowledge because that which can be done is capable of being otherwise, not art because action [*prāxis*] and making are different kinds of thing. It remains, then, that it is a true and reasoned state of capacity to act with regard to the things that are good or bad for man. For while making has an end other than itself, action cannot; for good action itself is its end" (*Nicomachean Ethics* 1140b).

troubles, through my own experience and to assign meaning to them in light of how my experience went. If a bunch of us are struggling with an obscure text, it seems to me that each of us is wrong in his own way, but each of those diverse ways of being wrong points to something that is missing in how we get at the text, and if we agree on this picture of the situation, and try to complement each other's lacks, then we might conjure a richer, if ultimately inconclusive, profile of the baffling, evanescent target of our efforts.

There is nothing right or wrong about any particular place where I land through this constant shifting, and I can certainly see myself, if my mood changed, landing somewhere else in similar circumstances (as I can see the corrupt politician, or my children, resisting the meanings I project onto them). Logic should not even try to tell you what is right or wrong, but rather what it means to be right or wrong. So should also, then, what I have done here: a logic of logics. It will tell you that the criteria of being right or wrong—specifically, of how to think, reason, and argue correctly or incorrectly—are not absolute, but relative to the logic you choose at any one time. And then if you, like me, prize the freedom allowed by this play among dissonant criteria, you will rejoice.

Coda: A Conversation Between Two Real Neighbors

Jack: Hi Don. How are you doing?
Don: Just the person I was looking for. I need to talk to you.
Jack: What about?
Don: It's about our backyards. See, it's been a while since we moved in...
Jack: A couple of months.
Don: Indeed, and so far we have been leaving everything open. Anyone can just wander back and forth between your property and mine.
Jack: Is there anything wrong with that?
Don: Yes, I think a lot is wrong with that. We should build a fence and share the expense.
Jack: That's some news. Why would I want to do that? I like the open space. I like the view. Why would I want to shut myself up in a cage?
Don: Well, to begin with, your dog always roams freely around here.
Jack: Immanuel is very friendly; he has never hurt anyone.
Don: Hasn't hurt me, or my family. But suppose we want to have company, a barbecue for a few friends, a party.
Jack: He wouldn't bother them either.
Don: Maybe not, but some people just don't like dogs. They are afraid, and dogs can feel that and become aggressive.
Jack: Immanuel would never react that way.
Don: Still, I would not want my guests to be troubled.
Jack: Even if you knew that there would be nothing to worry about?
Don: Yes, because *they* don't know it, or they wouldn't believe it, and that's enough. If they don't want your dog here, your dog shouldn't be here, end of story. There should be a fence that keeps him inside your property.
Jack: Just tell me when you're having company and I'll keep Immanuel inside the house. I don't see why we should pay for a major construction project because of something that will happen only once or twice a month.
Don: Well, it's not just that. There's also the issue of privacy.
Jack: What do you mean?

Don: In my own home, and in my backyard I should feel like I'm in my own home, I should be able to do whatever I please. Go around naked, if I want to, or make love in the grass. I shouldn't constantly have the impression that I am being watched by my neighbor.

Jack: I don't watch you, Don. I couldn't care less what you do.

Don: It doesn't matter if you watch me or not. What I said is that I should not *have the impression* that you are watching. And with all this open view I have that impression. I don't feel comfortable in my own home.

Jack: Don, what you are saying is ridiculous. See how our properties slope on this hill? See those other houses further up, no more than a hundred feet away? Even if you raise a fence to protect yourself from my eyes, what's to prevent those other people from watching? Are you going to build a fortress all around you? For all you know, anyone of them could be directing a telescope into your backyard and observing every detail of what happens there, even recording it for posterity if they wish, and there is nothing you could do about it because they would be on *their* property, doing what makes *them* feel at home.

Don: If they did that, I'd sue their asses. But why don't you let *me* worry about that and take care of it as I see fit, if and when I have a problem with it? I am trying to resolve an issue I have with you, and it's irrelevant whether I have issues with anyone else.

Jack: Alright, alright. So what about this? When you want to go around naked, or make love in the grass, you just send me a text that I should stay indoors, and I'll oblige.

Don: You are unreasonable! You prefer to shut yourself up, and your dog, just so that you can keep this space open. Why do you care so much about it?

Jack: People are different, we have to accept that. I enjoy the openness, the view, the sense of freedom it gives me. And I know that you will not always go around naked or make love in the grass; so most of the time I can be out here and even get to see you and have a nice conversation with you—not about a fence I hope.

Don: And because you want to enjoy the view and your sense of freedom I should be forced to send you a text every time I want you out of the way? What about *my* freedom? The freedom to forget that you even exist?

Jack: You are taking a very simplistic, myopic approach to this matter.

Don: How so?

Jack: Well, you think that being free is just acting as you please, following whatever whim strikes your fancy. But that is the freedom of a feather being blown by the wind, or rather by the *winds*, in the plural, prey to whatever little current of air plays with it, and pushes it this way or that. It is not the freedom of a rational being, who decides what he wants to do, based on his values and goals.

Don: Actually, Jack, it's you who are defending that simplistic position, not me.

Jack: What do you mean?

Coda: A Conversation Between Two Real Neighbors   153

Don: You are insisting on your freedom to wander around, to see what you like, much like primitive people, hunter-gatherers did in the old days, before there was any society to speak of, before there were any laws, any personal property. Then people evolved, and *freedom* evolved. It turned out that people are more free, not less, when they restrict their original capacity to wander aimlessly and accept limitations in the spaces that belong to them, when they restrict their original capacity to make, at every moment, an arbitrary choice and subject their choices to agreements with their peers. Their freedom can best express itself only by incorporating these limitations, because now there are more things they in fact *can* do—including doing whatever they do in a safer environment, where it can be done better.

Jack: I sort of figured that you would get there. Now you are going to tell me that civilized humans have a more mature freedom because they are integrated in a system of laws and a neat division of personal property that allows them to use what is uncontroversially their own in developing their own tastes and talents.

Don: That's what I would tell you if you didn't already tell it yourself.

Jack: And, presumably, you are also going to tell me that each of us is going to be freer if we just as neatly divide our properties. That my primitive sense of openness must evolve in an openness to be found beyond the building of a fence, not before it.

Don: That's right. Why do you think people say that good fences make good neighbors? Because all enjoy their homes more when their spaces are identified clearly; because it is easier to have healthy relations across a fence, free to choose when to make contact and when not to. Because that is what elementary, indefinite openness must graduate into: clarity. Just as a piece of writing is more open, it communicates more, when it abides by all the rules of spelling, and grammar, and syntax. To begin with, you may find those rules constraining—you certainly did as a child, when you were forced to learn them—but that very constraint is what ultimately frees your prose, frees you to write what you want to write, and to be understood by others. Even to be better understood by yourself.

Jack: Everything you said makes a lot of sense, so much that, as you have seen, I could say most of it myself, once I got the gist of it. I can do that because I have been there, because it is a path I have traveled, and in the houses I had before I was the one who went out and made the same speeches to my neighbors.

Don: Then we agree?

Jack: Yes and no. I agree that you are on the right track, but I believe you are stopping too soon.

Don: Now it's you who has to tell me what you mean.

Jack: I'll try. What you have done is show that things can turn into their opposites while still remaining themselves, indeed a better, more advanced form of themselves. A man who was short and weak as a child may turn into an adult that is no longer short and weak. A freedom that amounts to not

	knowing any rules or restrictions may turn into one that has accepted rules and restrictions, and that is *more* freedom because of that.
Don:	Yes; this is what I have done. What's wrong with it?
Jack:	Nothing is wrong. It is perfectly right, as far as it goes. Except that it doesn't quite go far enough. For the adult who is no longer short and weak is about ready to turn into an old man who is once more short and weak, though in a different way.
Don:	And freedom?
Jack:	Same thing, though it takes some effort, some ingenuity to see it. When you see it, on the other hand, it strikes you as the most natural thing in the world, the one you should have seen all along.
Don:	Well, make me see it, then.
Jack:	When you told your story, you ran together two distinct sorts of constraints: physical ones like fences (that supposedly make good neighbors) and legal ones like rules. So let me pick up the story where you left it off, and take it one or two stages further, so that you can see what I see now. In the old days of civilized humans, long after the hunter-gatherer era you were referring to, physical barriers were predominant. People locked themselves in their fortresses, worried about venturing into alien territory, armed themselves to the teeth when they had to.
Don:	Some still do.
Jack:	Yes, and why? Because they don't trust the institutions, so they revert to the past. But they cannot live in the past; they must get on with the present, with the new ways things are turning out.
Don:	And if they don't?
Jack:	They will be left behind. They will be the pitiful leftovers of a bygone time.
Don:	I don't know that they care.
Jack:	That's *why* they are pitiful leftovers. But forget about them. My point is that eventually the constraints, the "cages" that proved more liberating were found to be the legal ones, since they allowed a new form of wandering around, at the same time unimpeded and secure.
Don:	What are you talking about?
Jack:	I am talking about all the public spaces that have opened up as civilization progressed. I am talking about parks, roads, squares, sidewalks. All places where people are restrained not by physical barriers but by the respect for each other that the laws require. So that they can go and stay wherever they please, like those cavemen you were talking about, without being afraid of others doing violence to them.
Don:	Sometimes they do.
Jack:	Sure: a growth process is not painless, and is not victimless. But it is also unstoppable.
Don:	You are quite the optimist.
Jack:	It's not my personal sentiments that matter. It is the logic of evolution itself, as much of social evolution as of the biological variety.

Coda: A Conversation Between Two Real Neighbors 155

Don: I'm fine with everything you said; but I still don't see how it relates to our situation. We are not talking about public spaces here; we are talking about private ones, about my property and yours.

Jack: That's what I thought until a few years ago, and that's why I did have fences in those other houses. But then I realized that I had to go one step further. Public spaces are constituted by having everyone agree to leave them public, so there is no reason why two people like you and me could not agree to leave part of our spaces public, thus obtaining the same combination of liberty and security that people have on streets and sidewalks.

Don: I guess that pretty soon you'd want to make the insides of our houses into public spaces too.

Jack: Why not? That might be the next step. But progress is made one step at a time, so for the moment let's just enjoy the new phase of freedom our open backyards give us.

Don: I don't enjoy any of it. I don't enjoy your dog getting on this side and scaring my guests; I don't enjoy you watching me whatever I am doing; I don't enjoy the fact that someone might rob your house and then, finding no barrier between us, rob mine as well.

Jack: That's exactly as it should be. You should not enjoy any of that. Didn't I tell you that there are going to be growing pains?

Don: Who decides that these are *growing* pains? Why couldn't it be instead that *you* have been left behind by history, that you are a pitiful leftover overwhelmed by nostalgia, craving for a primitive sort of freedom? Why shouldn't *I* push you along, make you get on with the present, with how things are turning out?

Jack: So you didn't find my explanation irresistible, unobjectionable?

Don: Not at all. Have you been at the zoo, lately?

Jack: Not for a while. But what does the zoo have to do with anything?

Don: Even if it's been a while, you will remember the aviary.

Jack: How could I forget that? It's the most beautiful exhibit there: all sorts of greenery and rocks and waterfalls, not to mention all the birds flying here and there, or perching on a branch, birds of incredible colors, some with the strangest heads and feathers and beaks.

Don: I would imagine that the birds feel as if they were free to fly, or to perch, anywhere?

Jack: Yes, I would think so.

Don: But in fact they are in a cage.

Jack: I see what you mean. The aviary is enclosed by a plastic dome, but that is so high up and so large that the birds will never, or at least very rarely, become aware of it.

Don: I am sure you are right, and I am sure that, if and when they do, it won't bother them too much. They will think of it as just another obstacle in their path, and simply change directions.

Jack: Once I had this bird that for months was trying to get in through one of my windows, and went on forever pecking at it. The poor thing could not understand how something as transparent as air—something that for it *was* air—would not give in to its flight as air does. Eventually, it changed directions.

Don: OK, Jack. But that's birds. I just said that they will "think" of it, but that was not the right word to use. Birds don't think; so they are in a cage but do not realize it. Humans, however...

Jack: ... are thoughtful, reflective beings. So what?

Don: So, when they are in a cage, they will come to realize it.

Jack: Yes: if I were to lock you in an aviary, sooner or later you would become aware of the plastic dome and realize that you are caged in. I still don't see where you are going with this.

Don: You will soon enough. So you judge yourself free because you can see my plants, and can see me when I come out in my shorts and lie in the sun?

Jack: It's a beautiful sight, let me tell you.

Don: I am glad you like it. But, from where you are, you must also see the fence I have on the other side of my property, blocking your view from penetrating my other neighbor's property.

Jack: Yes, you guys sorted out that issue very quickly.

Don: He was not as eager as you are to look at me in my shorts. But the point is: though you don't have a wall here, you have one there. And you have a wall on the other side of *your* property, blocking your view of the street.

Jack: The city put that in.

Don: Nothing to argue there. And you have walls and fences everywhere else you look in the street. What you can actually see is very limited.

Jack: I can still see a lot. I can still walk everywhere.

Don: It depends on how you think of it. You can think of yourself as being able to go everywhere and see everything, *except for* some areas that are blocked, as they belong to someone else who has decided to wall himself in. *Or* you can think of yourself as locked into a particular path: forced to walk on it only and to see only what is accessible from it.

Jack: Yes, I agree that I can think of myself in either of those two ways, though I would add that the second way of thinking is far more depressing than the first one, and I would not want to indulge in it. But what about us now?

Don: Well, what difference does it make to you whether we build a fence between our properties, when you are caged anyway?

Jack: It makes a lot of difference, because without this fence of yours I can see a lot *more*, and feel a lot freer.

Don: And how much should you trust that feeling? Do you believe that someone sentenced to life in prison would be better off if he was given a larger cell, even one that was ten times larger?

Jack: He would be more comfortable.

Don: I don't deny it. But what I am asking is: would he be better off *as far as his freedom is concerned*?

Jack: Probably not.

Coda: A Conversation Between Two Real Neighbors 157

Don: That's right. Do you think it makes any difference for Truman Burbank, of *The Truman Show*, when he finally discovers his plastic dome, how large his cage is or how comfortable he has been living in it?

Jack: OK, I get it. But, given that I am caged no matter what, I might as well have a more spacious and comfortable cell.

Don: We will get to that. For the moment, let me just say that, as you are always caged, you are also always free.

Jack: You are all twists and turns. You've got to explain this now.

Don: It's not so hard, because again it's all a matter of how you think of it. It's just as I said before: you may think of all the movement you are allowed in this world as enclosed within fences and walls, as constrained by them no less than the walls of a jail constrain what a prisoner can (and, mostly, *cannot*) do; or you may think of all the fences and walls as rules that define your free movement, conditions of it, and think of your movement as remaining free and totally under your control within the space you are allowed, while abiding by those rules and conditions. For you are free to walk one step, and another one, and another one yet; and when you come before a wall you may think of yourself as blocked by the wall or, instead, as free to move laterally, away from it. None of this is going to change, no matter how far the wall is. It's always going to be dependent on how you think of it, on how you think of yourself.

Jack: So the prisoner can also think of himself as free?

Don: Of course, and yet his condition, whether he sees it as free or caged, can also be more or less convenient, more or less comfortable. But at least we've got the whole issue of freedom out of the way.

Jack: Let us talk about convenience and comfort, then, for the fence doesn't promise any of either to me.

Don: You said that one way of thinking of this is far more depressing than the other; so, since whether you are free or not depends on how you see yourself, you might as well "indulge" in seeing yourself as free and determine what is the best way of organizing your free space.

Jack: The best way to arrange my cell.

Don: If you prefer. In your own house, you find it more convenient and comfortable to have separate rooms, where you can attend separately to the various tasks of your everyday life, in ways that are more efficient and also more—how shall I put it?—"appropriate," rather than having, say, a single gigantic space, like a big tent, where all different activities are performed together: cooking and sleeping and listening to music and taking a shower and defecating.

Jack: That would be peculiar.

Don: It would, wouldn't it? Same here. Forgetting about freedom, which is a red herring, let us ask ourselves a factual question: is it more convenient and comfortable to leave all this space open, or wouldn't it be better to divide it up according to our different interests, so that you can visit when you want

	to but also do not have to stand my presence, or I yours, when we have other business to take care of?
Jack:	And the answer to this rhetorical question is?
Don:	But, clearly, it's better to divide up the space, so your dog can be better controlled...
Jack:	We are back at the dog; you are obsessed by the dog. You are annoying me.
Don:	No, Jack: *you* are annoying *me*. *I* feel caged, constrained by your presence all the time, all 365 days; *I* feel that it is an imposition on me to have to inform you when I want some privacy. I told you: I don't want to be forced to think about you all the time. If I care to see you and talk to you, I will come to your door and ring your bell; if I need your help I will text you. But most of the time I just want to be left alone, with you totally out of the picture.
Jack:	As I said, Don, people are different. You clearly feel one way and I feel another, and our feelings are irreconcilable; we feel contrary things.
Don:	But don't you see that my feelings should count more?
Jack:	Now, that's great. Why would that be?
Don:	Because the feelings that count the most are *hurt* feelings. Pain matters more than pleasure. Suppose someone takes pleasure in torturing another, and when the other guy tells him to stop he says: we are different; I would feel violated if you denied me this form of expression as much as you feel violated by the pain I inflict on you. Would that make any sense? Do you think there is any parity between these two contrary feelings? Should they be treated the same?
Jack:	No, they should not. But who is the torturer here?
Don:	You are. You are telling me that I should suffer my anxiety and my exasperation in silence so that you can have the pleasure of waking up to an open view.
Jack:	But don't you see that your description could be reversed? Then it would be my pain of waking up in a cage that counts less, for you, than your pleasure of finally being relieved of my presence!
Don:	Just like the torturer. He could reverse the description too, and say that his pain in not being able to express himself should count more than the pleasure his victim would feel in being relieved of his presence.
Jack:	But that would be absurd!
Don:	No more absurd than your own redescription.
Jack:	Wait a minute. The torturer is inflicting *physical* pain, is literally destroying the body of his victim. The pain he would feel if they made him stop, even admitting that we can allow such talk, would only be psychological; so clearly there is no parity. But we are *both* talking about a psychological condition.
Don:	Sure, Jack, but there are also *psychological* torturers. People who put others in double-bind situations, with no way out, no viable option for how to behave, and end up destroying them to an extent even more radical and irremediable than many physical torturers.

Coda: A Conversation Between Two Real Neighbors

Jack: So I would be torturing you psychologically?
Don: That's what I have been saying.
Jack: And your proposal for how to remedy that is to torture *me* psychologically.
Don: I don't see it that way.
Jack: I know that you don't see it that way, but it is your word against mine. So all you are saying is that your pain should count more because your word counts more.
Don: I see that there is no way to resolve this.
Jack: No, there isn't. But I cannot prevent you from raising all the fences you want on your own property. I won't like it, just as I wouldn't if you planted a tree that blocked the view; but I can't decide for you. So go ahead if that's what you want; but be sure that you stay clear of everything that is mine. And be sure that I won't pay a penny for it.
Don: If I had been happy to do what you just said, I would not have needed to talk to you, and I could have avoided all this aggravation.
Jack: Yes, you could have, and I could have too. I could have relaxed down there in the shade instead of listening to your complaints. We have been wasting our time. Rather, *you* have been wasting *my* time.
Don: I don't think so. Now that we have agreed that there will be a fence, we must get to the sharing of the costs of it.
Jack: There will be no sharing, I told you. You do it on your own in your own backyard.
Don: I don't think that's fair.
Jack: Fair? You want to force me to do something I don't want and I don't like, and want me to pay for it, and I am not being fair?
Don: Sure, because once the fence is up you will profit from it.
Jack: I will hate every inch of it!
Don: You say so, but it may be a tactic for having me take care of the whole project. Once it's up, you will enjoy privacy and security as much as I will, at my expense.
Jack: I told you I don't care about privacy and security—about *what you call* privacy and security.
Don: Yes, you told me. But the privacy and security the fence will provide you are objective goods, and for all I know might even increase the value of your house. That you say that they have no value does not make it so.
Jack: They have no value for me.
Don: Even if I give you the benefit of the doubt, and believe that you not playing games with me, all I can grant you is that they have no value for you *now*. Tomorrow, or next month, you will see the light, or you will sell the house, *at a higher price*, and the new owner will see the light, and appreciate the privacy and security the fence gives him, and I will have been paying for it.
Jack: OK, so suppose I agree that privacy and security are objective values. Do you agree that openness and freedom are also objective values? And that I have a right to set my own priorities among all these values? That I can

	decide to value freedom and openness *more* than your damned privacy and security?
Don:	There must be an objective way of settling this dispute. There must be experts who can determine how to rank these values objectively.
Jack:	Yes, and there must be laws against the kind of harassment you have been subjecting me to. There must be ways of protecting an honest citizen from this invasion of privacy—of *my* privacy, that you seem to have no respect for.
Don:	I have a right to interrupt your lazy contemplation of openness to bring up an important matter.
Jack:	I don't find this matter at all important. I find it petty, not to mention intrusive.
Don:	That's another thing we see differently. But let me give it another shot. You have agreed that privacy and security are objective values, and I may agree that freedom and openness are too.
Jack:	Now that's making progress!
Don:	You also said that you rank freedom and openness higher, but perhaps you can tell me *how much* higher. Do you think that freedom and openness are worth twice as much as privacy and security, or three times?
Jack:	A hundred times!
Don:	Don't be foolish. Give me a reasonable figure.
Jack:	Ten times.
Don:	Then what if I asked you to pay for one tenth of the fence, and to build one tenth of it on your property?
Jack:	I would say, No.
Don:	But why?
Jack:	Because *you* should be compensating me for the loss of something that I value so much more!
Don:	So no compromise is possible.
Jack:	No. Build whatever cage you want, on your property, and leave me alone.
Don:	You will be hearing from my lawyers.
Jack:	*My* lawyers will.

Bibliography[1]

Adorno, Theodor Wiesengrund. "The Stars Down to Earth: The *Los Angeles Times* Astrology Column." *Telos* 19 (1974), pp. 13–90.
Anselm. *Monologion and Proslogion with the Replies of Gaunilo and Anselm*, translated by Thomas Williams. Indianapolis (IN): Hackett, 1995.
Aristophanes. *Clouds*, translated by Peter Meineck. Indianapolis (IN): Hackett, 2000.
Aristotle. *The Complete Works*, in two volumes, edited by Jonathan Barnes. Princeton: Princeton University Press, 1984.
Armstrong, Karen. *The Case for God*. New York: Knopf, 2009.
Barthes, Roland. *The Fashion System*, translated by Matthew Ward and Richard Howard. New York: Farrar, Straus and Giroux, 1983.
Bencivenga, Ermanno. "A Semantics for a Weak Free Logic." *Notre Dame Journal of Formal Logic* 19 (1978), pp. 646–652.
Bencivenga, Ermanno. *A Theory of Language and Mind*. Berkeley (CA): University of California Press, 1997.
Bencivenga, Ermanno. *Ethics Vindicated: Kant's Transcendental Legitimation of Moral Discourse*. New York: Oxford University Press, 2007.
Bencivenga, Ermanno. *Hegel's Dialectical Logic*. New York: Oxford University Press, 2000.
Bencivenga, Ermanno. *Kant's Copernican Revolution*. New York, Oxford University Press, 1987.
Bencivenga, Ermanno. "On Good and Bad Arguments." *Journal of Philosophical Logic* 8 (1979), pp. 247–259.
Bencivenga, Ermanno. "On the Very Possibility of a Formal Logic and Why Dialectical Logic Cannot be One." *The Philosophical Forum* 46 (2015), pp. 275–286.
Bencivenga, Ermanno. *Return from Exile: A Theory of Possibility*. Lanham (MD): Lexington Books, 2013.
Bencivenga, Ermanno. "What Is Logic About?" *European Review of Philosophy* 4 (1999), pp. 5–19.
Benjamin, Walter. "The Work of Art in the Age of Mechanical Reproduction," translated by Harry Zohn. In *Illuminations: Essays and Reflections*. New York: Schocken Books, 1969, pp. 217–251.

---

[1] Note: References to works by Plato and Aristotle in this book are to the Stephanus and Bekker editions, respectively, though, when a quote is given, the translation being used is the one indicated below. References to Kant's *Critique of Pure Reason* are given by using the A/B numbers. All other references are to the paginations in the editions listed below. I have mostly uniformed quotes to American spelling.

Bergson, Henri. *Creative Evolution*, translated by Arthur Mitchell. London: Macmillan, 1911. (1911c).
Bergson, Henri. *Laughter: An Essay on the Meaning of Comic*, translated by Cloudesley Brereton and Fred Rothwell. London: Macmillan, 1911. (1911b).
Bergson, Henri. *Matter and Memory*, translated by Nancy Margaret Paul and William Scott Palmer. London: Swan Sonnenschein, 1911. (1911a).
Bergson, Henri. *The Two Sources of Morality and Religion*, translated by Ruth Ashley Audra and Cloudesley Brereton, with the assistance of William Horsfall Carter. London: Macmillan, 1935.
Bergson, Henri. *Time and Free Will: An Essay on the Immediate Data of Consciousness*, translated by Frank Lubecki Pogson. London: Allen & Unwin, 1910.
Brandom, Robert Boyce. "A Binary Sheffer Operator Which Does the Work of Quantifiers and Sentential Connectives." *Notre Dame Journal of Formal Logic* 20 (1979), pp. 262–264.
Bruno, Giordano. *The Ash Wednesday Supper*, translated by Edward Gosselin and Lawrence Lerner. Toronto: University of Toronto Press, 1995.
Calvino, Italo. *The Uses of Literature: Essays*, translated by Patrick Creagh. San Diego (CA): Harcourt Brace Jovanovich, 1986.
Campanella, Tommaso. *Metafisica*, edited by Paolo Ponzio. Bari: Levante, 1994.
Cohen, Paul Joseph. *Set Theory and the Continuum Hypothesis*. New York: W. A. Benjamin, 1966.
Croce, Benedetto. *Filosofia e storiografia*. Napoli: Bibliopolis, 2005.
Darwin, Charles. *The Origin of Species*. Oxford: Oxford University Press, 1996.
Deleuze, Gilles. *The Logic of Sense*, translated by Mark Lester with Charles Stivale. New York: Columbia University Press, 1990.
Derrida, Jacques. *Of Grammatology*, translated by Gayatri Chakravorty Spivak. Baltimore (MD): Johns Hopkins University Press, 1974.
Derrida, Jacques. *Writing and Difference*, translated by Alan Bass. Chicago: University of Chicago Press, 1978.
Descartes, René. *Meditations on First Philosophy, with Objections and Replies*. In *The Philosophical Writings of Descartes*, translated by John Cottingham, Robert Stoothoff, and Dugald Murdoch. Cambridge: Cambridge University Press, 1984/1985, vol. II, pp. 3–62, 66–383.
Descartes, René. *Principles of Philosophy*. In Cottingham *et al.* cit., vol. I, pp. 179–291.
Descartes, René. *Rules for the Direction of the Mind*. In Cottingham *et al.* cit., vol. I, pp. 9–76.
Frege, Gottlob. *The Foundations of Arithmetic*, translated by John Langshaw Austin. Evanston (IL): Northwestern University Press, 1980.
Frege, Gottlob. *Translations from the Philosophical Writings of Gottlob Frege*, edited by Peter Geach and Max Black. Oxford: Blackwell, 1952.
Frege, Gottlob. *Collected Papers on Mathematics, Logic, and Philosophy*, edited by Brian McGuinness. Oxford: Blackwell, 1984.
Frege, Gottlob. *Collected Papers on Mathematics, Logic, and Philosophy*, edited by Brian McGuinness. Hoboken (NJ): Wiley, 1991.
Galilei, Galileo. *Dialogue concerning the Two Chief World Systems*, translated by Stillman Drake. New York: Modern Library, 2001.
Gauthier, David. *Morals by Agreement*. Oxford: Clarendon Press, 1986.
Hegel, Georg Wilhelm Friedrich. *Elements of the Philosophy of Right*, translated by Hugh Barr Nisbet. Cambridge: Cambridge University Press, 1991. (1991b).
Hegel, Georg Wilhelm Friedrich. *Faith and Knowledge*, translated by Walter Cerf and Henry Silton Harris. Albany (NY): State University of New York Press, 1977. (1977a).
Hegel, Georg Wilhelm Friedrich. *Lectures on the History of Philosophy*, in three volumes, translated by Elizabeth Sanderson Haldane and Frances Helen Simson. Lincoln (NE): University of Nebraska Press, 1995.

Hegel, Georg Wilhelm Friedrich. *Lectures on the Philosophy of Religion*, in three volumes, translated by Robert F. Brown, Peter Crafts Hodgson, and J. Michael Stewart, with the assistance of Henry Silton Harris. Berkeley (CA): University of California Press, 1984/1988.

Hegel, Georg Wilhelm Friedrich. *Phenomenology of Spirit*, translated by Arnold Vincent Miller. Oxford: Oxford University Press, 1977. (1977b).

Hegel, Georg Wilhelm Friedrich. *Philosophy of History*, translated by John Sibree. Buffalo (NY): Prometheus Books, 1991. (1991c).

Hegel, Georg Wilhelm Friedrich. *Philosophy of Mind*, translated by Arnold Vincent Miller. Oxford: Clarendon Press, 1971.

Hegel, Georg Wilhelm Friedrich. *Science of Logic*, translated by Arnold Vincent Miller. Atlantic Highlands (NJ): Humanities Press, 1990.

Hegel, Georg Wilhelm Friedrich. *The Encyclopaedia Logic: Part I of the Encyclopaedia of Philosophical Sciences*, translated by Théodore F. Geraets, Wallis Arthur Suchting, and Henry Silton Harris. Indianapolis (IN): Hackett Publishing Company, 1991 (1991a).

Hegel, Georg Wilhelm Friedrich. *The Letters*, translated by Clark Butler and Christiane Seiler. Bloomington: Indiana University Press, 1984.

Heidegger, Martin. *Being and Time*, translated by John Macquarrie and Edward Robinson. New York: Harper & Row, 1962.

Henkin, Leon. "The Completeness of the First-Order Functional Calculus." *Journal of Symbolic Logic* 14 (1949), pp. 159–166.

Hilbert, David. *Grundlagen der Geometrie*. Leipzig: Teubner, 1903.

Hilbert, David. "On the Infinite," translated by Stefan Bauer-Mengelberg. In *From Frege to Gödel: A Source Book in Mathematical Logic, 1879–1931*, edited by Jean van Heijenoort. Cambridge (MA): Harvard University Press, 1967, pp. 369–392.

Hobbes, Thomas. *The English Works of Thomas Hobbes of Malmesbury*, vol. I, edited by William Molesworth. London: John Bohn, 1839.

Kant, Immanuel. *Critique of Pure Reason*, translated by Paul Guyer and Allen William Wood. Cambridge: Cambridge University Press, 1998.

Kant, Immanuel. *Critique of the Power of Judgment*, translated by Paul Guyer and Eric Matthews. Cambridge: Cambridge University Press, 2000.

Kant, Immanuel. *Metaphysical Foundations of Natural Science*, translated by Michael Friedman. Cambridge: Cambridge University Press, 2004.

Kirk, Geoffrey Stephen, John Earle Raven, and Malcolm Schofield. *The Presocratic Philosophers*. Cambridge: Cambridge University Press, 1983.

Kuhn, Thomas Samuel. *The Structure of Scientific Revolutions*. Chicago: University of Chicago Press, 1962.

Leibniz, Gottfried Wilhelm. *Die philosophischen Schriften*, vol. 4, edited by Karl Immanuel Gerhardt. Berlin: Weidmann, 1880.

Löwenheim, Leopold. "On Possibilities in the Calculus of Relatives," translated by Stefan Bauer-Mengelberg. In van Heijenoort cit., pp. 232–251.

Peano, Giuseppe. *Arithmetices principia nova methodo exposita*. Torino: Bocca, 1889.

Plato. *Republic*, translated by Robin Waterfield. Oxford: Oxford University Press, 1993.

Priest, Graham, Richard Routley, and Jean Norman (editors). *Paraconsistent Logic: Essays on the Inconsistent*. Munich: Philosophia Verlag, 1990.

Putnam, Hilary. "Models and Reality." *Journal of Symbolic Logic* 45 (1980), pp. 464–482.

Ramsey, Frank Plumpton. *Philosophical Papers*, edited by David Hugh Mellor. Cambridge: Cambridge University Press, 1990.

Reid, Constance. *Hilbert*. Göttingen: Copernicus, 1996.

Russell, Bertrand. *Our Knowledge of the External World*. London: Allen & Unwin, 1914.

Sainsbury, Mark. "Concepts without Boundaries." In *Vagueness: A Reader*, edited by Rosanna Keefe and Peter Smith. Cambridge (MA): MIT Press, 1996, pp. 251–264.

Sartre, Jean-Paul. *Being and Nothingness*, translated by Hazel Estella Barnes. New York: Washington Square Press, 1992.

Skolem, Thoralf. "Logico-combinatorial Investigations in the Satisfiability or Provabilitiy of Mathematical Propositions: A Simplified Proof of a Theorem by L. Löwenheim and Generalizations of the Theorem," translated by Stefan Bauer-Mengelberg. In van Heijenoort cit., pp. 254–263 (1967a).

Skolem, Thoralf. "The Foundations of Elementary Arithmetic Established by Means of the Recursive Mode of Thought, without the Use of Apparent Variables Ranging over Infinite Domains," translated by Stefan Bauer-Mengelberg. In van Heijenoort cit., pp. 303–333 (1967b).

Spinoza, Benedict de. *Ethics*, translated by Edwin Curley. London: Penguin, 1996.

Strawson, Peter Frederick. *Introduction to Logical Theory*. London: Methuen, 1952.

Vico, Giambattista. *The New Science*, translated by Thomas Goddard Bergin and Max Harold Fisch. Ithaca: Cornell University Press, 1984.

Weyl, Hermann. *The Continuum: A Critical Examination of the Foundation of Analysis*, translated by Stephen Pollard and Thomas Bole. Kirksville (MO): Thomas Jefferson University Press, 1987.

Wittgenstein, Ludwig. *Philosophical Remarks*, edited by Rush Rhees, translated by Raymond Hargreaves and Roger White. Oxford: Basil Blackwell, 1975.

Wittgenstein, Ludwig. *Tractatus Logico-Philosophicus*, translated by David Pears and Brian McGuinness. London: Routledge and Kegan Paul, 1961.